Alexander Wiedenmann

Glimpse into an ATP synthase's F0 motor at work

Alexander Wiedenmann

Glimpse into an ATP synthase's F0 motor at work

Requirements for torque generation in proton and sodium dependent F-type ATP synthases

Südwestdeutscher Verlag für Hochschulschriften

Impressum/Imprint (nur für Deutschland/ only for Germany)
Bibliografische Information der Deutschen Nationalbibliothek: Die Deutsche Nationalbibliothek verzeichnet diese Publikation in der Deutschen Nationalbibliografie; detaillierte bibliografische Daten sind im Internet über http://dnb.d-nb.de abrufbar.
Alle in diesem Buch genannten Marken und Produktnamen unterliegen warenzeichen-, marken- oder patentrechtlichem Schutz bzw. sind Warenzeichen oder eingetragene Warenzeichen der jeweiligen Inhaber. Die Wiedergabe von Marken, Produktnamen, Gebrauchsnamen, Handelsnamen, Warenbezeichnungen u.s.w. in diesem Werk berechtigt auch ohne besondere Kennzeichnung nicht zu der Annahme, dass solche Namen im Sinne der Warenzeichen- und Markenschutzgesetzgebung als frei zu betrachten wären und daher von jedermann benutzt werden dürften.

Verlag: Südwestdeutscher Verlag für Hochschulschriften Aktiengesellschaft & Co. KG
Dudweiler Landstr. 99, 66123 Saarbrücken, Deutschland
Telefon +49 681 37 20 271-1, Telefax +49 681 37 20 271-0, Email: info@svh-verlag.de
Zugl.: Zürich, ETH, Diss., 2008

Herstellung in Deutschland:
Schaltungsdienst Lange o.H.G., Berlin
Books on Demand GmbH, Norderstedt
Reha GmbH, Saarbrücken
Amazon Distribution GmbH, Leipzig
ISBN: 978-3-8381-0659-5

Imprint (only for USA, GB)
Bibliographic information published by the Deutsche Nationalbibliothek: The Deutsche Nationalbibliothek lists this publication in the Deutsche Nationalbibliografie; detailed bibliographic data are available in the Internet at http://dnb.d-nb.de.
Any brand names and product names mentioned in this book are subject to trademark, brand or patent protection and are trademarks or registered trademarks of their respective holders. The use of brand names, product names, common names, trade names, product descriptions etc. even without a particular marking in this works is in no way to be construed to mean that such names may be regarded as unrestricted in respect of trademark and brand protection legislation and could thus be used by anyone.

Publisher:
Südwestdeutscher Verlag für Hochschulschriften Aktiengesellschaft & Co. KG
Dudweiler Landstr. 99, 66123 Saarbrücken, Germany
Phone +49 681 37 20 271-1, Fax +49 681 37 20 271-0, Email: info@svh-verlag.de

Copyright © 2009 by the author and Südwestdeutscher Verlag für Hochschulschriften Aktiengesellschaft & Co. KG and licensors
All rights reserved. Saarbrücken 2009

Printed in the U.S.A.
Printed in the U.K. by (see last page)
ISBN: 978-3-8381-0659-5

Mit dem Wissen wächst der Zweifel.
J. W. von Goethe

Für Eva, Jacob und Lucas

Acknowledgements

This thesis was only possible with the help and support of many people.

First, I would like to thank Prof. Peter Dimroth for giving me the opportunity to work on the world's most fascinating protein complex. Moreover, I would like to express my gratitude for the freedom, fairness and the scientific input he provided throughout my time in his lab.

I further would like to thank Prof. Markus Aebi for taking over the official supervision of my thesis.

I am indebted to Christoph von Ballmoos for his guidance, his motivation and his help throughout the 4 years. Though the times we experienced were not always easy I fear it will be hard to find such an enjoyable and fruitful collaboration again.

During my thesis I was fortunate enough to have swift *clone managers* besides me, namely Carole and Judith, who did all the cloning and saved me a lot of work.

I was fortunate enough to supervise the diploma work of Damien *Skaterboy* Morger which was a very enjoyable task due to his his enthusiasm and devotion (we do not talk about his sense of order here....).

Prof. Gregory Cook I would like to thank for many valuable discussions, fruitful ideas, motivation and a very effective letter of recommendation.

A number of people in the lab helped me to overcome the obstacles I had to tackle especially during the first two years. I would like to thank especially Alexandra, Christian, Martin Poos and Yolanda.

I also want to mention Markus Seeger for the time of our presidency of the *VIM*, the discussions, beers and pizzas.

Of course I want to thank all members of the Dimroth group for enjoyable times in the lab. The board and staff of the institute I want to thank for providing an excellent and highly efficient infrastructure for research.

I would also like to mention Prof. Lutz Schmitt and Jelena Zaitseva who both gave me a solid knowledge in biochemical research during my diploma thesis and were always a reliable reference during my PhD thesis.

Last but not least the successful completion of my thesis was only possible with the support and many sacrifices from Eva, Jacob and my Mum.

Contents

Zusammenfassung		1
Summary		3
1 General Introduction		**5**
1.1	ATP - the universal currency of free energy in biological systems	5
1.2	F-type ATPases	6
1.3	The F_1 part is responsible for ATP hydrolysis and synthesis	10
1.4	The F_0 part couples ion transport and torque generation	14
	1.4.1 Subunit a	15
	1.4.2 Subunit b	16
	1.4.3 Subunit c	17
1.5	Na^+-dependent F_1F_0 ATP synthases	20
	1.5.1 *Propionigenium modestum* and *Ilyobacter tartaricus*	20
1.6	The ion pathway through the membrane	21
1.7	Driving forces for F-type ATPases	22
1.8	Model of the F_0 motor	24
1.9	Aim of this work	25
2 $\Delta\psi$ and ΔpH are equivalent driving forces for the F_0 part		**27**
2.1	Abstract	27
2.2	Introduction	27
2.3	Material and Methods	28
2.4	Results and Discussion	32
	2.4.1 Unidirectional Reconstitution	32
	2.4.2 Pyranine as indicator of internal pH change	33
	2.4.3 Initial rates of H^+-transport through F_0 of *E. coli* or spinach chloroplasts	37
	2.4.4 Number of functional F_0 molecules	40
	2.4.5 $\Delta\psi$- and ΔpH-driven H^+-transport in synthesis and hydrolysis direction	41

Contents

    2.4.6   Ohmic conductance of the *E. coli* F_0 part 43
    2.4.7   Inhibition of H^+-translocation by tributyltin chloride 43
2.5   Concluding remarks 48

3 Glimpse into an ATP synthase's F_0 motor at work   49
3.1   Abstract 49
3.2   Glimpse into an ATP synthase's F_0 motor at work 50
3.3   Methods summary 58
3.4   Supplementary Discussion 58
    3.4.1   Asymmetric Na^+ binding affinities of the F_0 ATP synthase of *P. modestum* 58
    3.4.2   Spectroscopic surveillance of critical parameters during ATP synthesis in inverted membrane vesicles of *E. coli* 59
    3.4.3   Molecular mechanism of the F_0-ATP synthase in different operation modes 60
    3.4.4   Final conclusions 63
3.5   Supplementary Methods 65
    3.5.1   Enzyme preparation and reconstitution into phospholipid vesicles 65
    3.5.2 Na^+-transport measurements 65
    3.5.3 H^+-transport measurements 65
    3.5.4   ATP synthesis measurements with *P. modestum* liposomes 65
    3.5.5   Reconstitution and ATP synthesis measurements with *E. coli* ATP synthase 66

4 Impact of ΔpNa on the F_0 part from *Propionigenium modestum*   67
4.1   Introduction 67
4.2   Material and Methods 69
    4.2.1   Enzyme purification and reconstitution 69
    4.2.2   Measurement of ATP hydrolysis activity 69
    4.2.3   Western Blot analysis 69
    4.2.4 Na^+-transport measurements 69
    4.2.5   Determination of ATP synthesis 70
4.3   Results and Discussion 71
    4.3.1   Orientation of *P. modestum* F_0 part in liposomes 71
    4.3.2 ΔpNa is an effective driving force in synthesis, but not hydrolysis direction 72
    4.3.3 ΔpNa does not affect $\Delta\psi$-driven Na^+-transport in hydrolysis direction . . 74
    4.3.4   The K_{app} for Na^+ during ATP synthesis is much higher than during ATP hydrolysis 74
    4.3.5 $\Delta\psi$ does not affect K_D during ATP synthesis 75

  4.3.6 A model of torque generation in the *P. modestum* F_0 part in response to $\Delta\psi$ and ΔpNa 76

5 General Discussion - Structure and Function of the a-Subunit in F-type ATPases 83
 5.1 Topology of the a-subunit 84
 5.2 Structural information provided by Cys-Cys cross-linking 87
 5.3 Cysteine scanning mutagenesis of the a-subunit 89
 5.4 Functional mutations in the *E. coli* a-subunit 92
  5.4.1 Essential residues in the ATP synthase 97
  5.4.2 Mutants leading to impaired growth on succinate 98
  5.4.3 Second site suppressor mutations 99
 5.5 Sequence alignment of bacterial, mitochondrial and chloroplast a-subunits 102
  5.5.1 Sequence variations of the triad 218/ 219/ 245 102
 5.6 Concept of lateral proton diffusion 103
 5.7 Outlook 104

Appendix I   **107**

Appendix II   **109**

Bibliography   **111**

Contents

Zusammenfassung

Adenosintriphosphat (ATP) ist der universelle Energieträger in der gesamten belebten Welt. Eine Vielzahl von zellulären Prozessen wie z.b. Vesikeltransport, Elektrolyt Homöostase und DNA Replikation sind von der Verfügbarkeit von ATP abhängig.

Das meiste ATP wird durch einen Enzymkomplex, die sogenannte F_1F_0 ATP Synthase, hergestellt. Dieser Komplex besteht aus dem Membran-assoziierten F_1 Teil und dem Membranintegralen F_0 Teil. Die ATP Synthase hat nicht nur eine zentrale Rolle im zellulären Stoffwechsel, sondern besitzt zusätzlich einen faszinierenden und einmaligen Mechanismus: Wie bei einem Mühlrad fliessen H^+- oder Na^+- Ionen entlang des elektrochemischen Gradienten der Zellmembran durch den F_0 Teil und erzeugen durch die dabei frei werdende Energie eine Rotation. Diese Rotation wir über eine zentrale, asymmetrische Welle vom F_0 Teil auf den F_1 Teil übertragen, wo die Energie schliesslich genutzt wird, um ADP und P_i zu ATP zu kondensieren.

Der elektrochemische Gradient, der die ATP Synthase antreibt, wird entweder von den Atmungskettenkomplexen oder durch die Reaktionszentren der Fotosynthese erzeugt. In seltenen Fällen wird die ATP Synthase durch einen Na^+-Gradient angetrieben, der von einer Membranständigen Decarboxylase erzeugt wird.

Während der F_1 Teil der ATP Synthase im Detail erforscht ist, liegen grosse Teile des Mechanismus und der Struktur des F_0 Teils noch im Dunkeln.

In der vorliegenden Arbeit sollten die Antriebkräfte untersucht werden, die zur Rotation im isolierten F_0 Teil bzw. zur ATP Synthese in der F_1F_0 ATP Synthase führen. Darüber hinaus sollten die Antriebskräfte zwischen dem H^+-gekoppelte Enzym von *E. coli* untersucht und mit der Na^+-gekoppelte ATP Synthase von *P. modestum* verglichen werden.

Um die Antriebskräfte definieren zu können, die zur Rotation im F_0 Teil führen, wurde ein Protokoll entwickelt, das zu einer einheitlichen Orientierung der ATP Synthase in Liposomen führt. Dadurch konnte der H^+-Transport in bzw. aus den Liposomen genau quantifiziert werden. Die Methode basiert auf dem Farbstoff Pyranin, einem wasserlöslichen Fluorophor, der seine optischen Eigenschaften mit der H^+-Konzentration ändert. Mit dieser Methode wurde der Einfluss von ΔpH und $\Delta \psi$ auf den F_0 Teil der ATP Synthasen von *E. coli* und aus Spinat Chloroplasten bestimmt (Kapitel 2).

Um die Antriebskräfte des F_0 Teils mit denen des Gesamtenzyms vergleichen zu können, wurden die Bedingungen untersucht, welche die Synthese von ATP mit den Enzymen von

E. coli und *P. modestum* ermöglichen. Ein Ionen-Konzentrationsgradient von $\sim$ 30 - 50 mV war in beiden Enzyme für die Synthese von ATP unerlässlich. Erstaunlicher war jedoch das Ergebnis, dass $\sim$ 100-fach höhere Ionenkonzentrationen für die Synthese von ATP benötigt wurden, verglichen mit der Hydrolyse von ATP. Zum ersten Mal wurde die Ionenkonzentration auf der periplasmatischen Seite als wichtige Voraussetzung für die ATP Synthese charakterisiert. Auf Grund dieser neuen Erkenntnisse wurde ein erweitertes Modell für die Erzeugung von Rotation im F_0 Teil der ATP Synthase entworfen.

Frühere Untersuchungen haben gezeigt, dass ein ΔpNa alleine nicht zum Na^+ Transport führt. Es war daher überraschend, dass ein ΔpNa für die ATP Synthese unerlässlich war (siehe Kapitel 3). Dieser scheinbare Widerspruch wurde in Kapitel 4 eingehend untersucht. Es zeigte sich, dass die Rekonstitutionsmethode für die *P. modestum* ATP Synthase ebenfalls zu einer einheitlichen Orientierung der ATP Synthase in die Liposomen führte. Daher wurde in den vorangegangenen Studien ausschliesslich Na^+ Transport in Synthese Richtung beobachtet. Wurden die Triebkräfte in der umgekehrten Orientierung angelegt, führten sowohl ein ΔpNa als auch ein $\Delta\psi$ sowie eine Kombination aus beiden Triebkräften zum Ausstrom von Na^+, was mit den Beobachtungen bei der ATP Synthese übereinstimmt. Die Ergebnisse zeigten ausserdem, dass ΔpNa und $\Delta\psi$ unterschiedliche Wirkungen auf den Transport in Hydrolyse oder Synthese Richtung haben.

Im letzten Kapitel (Kapitel 5) wurde eine Übersicht zu den Erkenntnissen über die a-Untereinheit der letzten 30 Jahre erstellt. Die Erkenntnisse und ihre Bedeutung aus den verschiedenen Studien wurden diskutiert und neue Aspekte zum Mechanismus der a-Untereinheit vorgeschlagen.

Summary

Adenosine triphosphate (ATP) is the universal carrier of free energy in the living world. A plethora of cellular reactions like vesicle transport, electrolyte homeostasis, and replication depend on the availability of ATP. The majority of ATP is generated by an enzyme complex termed F_1F_0 ATP synthase.

The bipartite enzyme, which consists of a membrane-embedded F_0 part and a water-soluble F_1 part, has not only a central role in cellular metabolism but in addition features a fascinating catalytic mechanism: Energy stored in the transmembrane electrochemical gradient is converted into rotation by an H^+ or Na^+ current through the membrane-embedded F_0 part, reminiscent of a waterwheel. The rotation is transmitted by the central stalk from the F_0 to the water-exposed F_1 part, where the rotation is converted into chemical energy by condensation of ADP and P_i to ATP.

The electrochemical gradient, which fuels the ATP synthase, is created by the respiratory chain, photosynthetic reaction centers or in some cases by a membrane-bound decarboxylase, which catalyzes Na^+-transport.

Compared to the thoroughly investigated F_1 part, knowledge about the mechanism of the F_0 part is still scarce. The present thesis aimed to elucidate the driving forces that are required for torque generation during ATP synthesis in the holoenzyme as well as in the isolated F_0 part. Furthermore, we designed experiments to compare the requirements for ATP synthesis in the Na^+-dependent and the H^+-dependent ATP synthases from *P. modestum* and *E. coli*, respectively.

To characterize the driving forces causing rotation in the F_0 part, a method was developed which led to unidirectional incorporation of the *E. coli* ATP synthase in preformed lipid vesicles. This allowed the accurate quantification of H^+-transport in and out of F_0-containing proteoliposomes. The technique is based on the hydrophilic fluorophore pyranine which changes its fluorescent properties in response to the proton concentration (pH). Using these techniques, we studied the impact of ΔpH and $\Delta\psi$ on H^+-transport through F_0 parts of the ATP synthase from spinach chloroplasts and *E. coli* (chapter 2).

In addition, the requirements for rotation during ATP synthesis in the holoenzyme were investigated in the H^+- and Na^+-dependent enzymes from *E. coli* and *P. modestum*. Although small (30 - 50 mV), an ion concentration gradient was indispensable for ATP synthesis in both enzymes. Much more surprising was the observation that $\sim$ 100 times higher Na^+- or H^+-concentrations were required for the synthesis than the hydrolysis of ATP. For the first time the ion concentration at the periplasmic side is recognized as critical determinant for ATP synthesis. On the basis of the gathered experimental evidence an advanced model for torque generation in the F_0 part of F-type ATPases was constructed. In earlier reports, ΔpNa was found to be insufficient to elicit rotation in the *P. modestum* F_0 part. The studies described in chapter 3, however, assigned an indispensable role to ΔpNa during ATP synthesis.

This apparent conundrum was investigated in more detail and is described in chapter 4. We inspected our reconstitution procedure and found predominantly uniformly oriented enzymes in the proteoliposomes. As a result, only transport in hydrolysis was monitored in the earlier studies. When we reversed the transport direction, ΔpNa, $\Delta\psi$ or a combination of both led to Na^+-transport, matching the requirements for ATP synthesis. The data demonstrated that the two types of driving forces (ΔpNa and $\Delta\psi$) have a different impact on the *P. modestum* F_0 part, whether it works in hydrolysis or synthesis mode.

To get a better picture of the a-subunit, an overview of investigations on the a-subunit during the last 30 years is given in the general discussion (chapter 5). The significance and usefulness of applied techniques are discussed and novel ideas about functional aspects of the ATP synthase are proposed.

1 General Introduction

1.1 ATP - the universal currency of free energy in biological systems

One hallmark of life is an active metabolism which is divided into anabolic and catabolic pathways. Anabolic reactions synthesize molecules from smaller units under the consumption of energy whereas catabolic reactions break down molecules into smaller units generating energy. Fritz Lipmann proposed in 1941 that the link between these two pathways is adenosine triphosphate (ATP). Today, ATP is recognized as the universal carrier of free energy throughout the living world. It has a key role in the development of life. Nature decided at one point to use ATP when free energy is required in biological systems even though completely equivalent compounds like GTP would have been available. The hydrolysis of the terminal phosphodiester bond from ATP to ADP liberates an energy of approx. 30 kJ/ mol and can change the reaction constant by a factor of 10^8 (Stryer 1995). A plethora of cellular reactions depend on the availability of ATP. ATP is required for many energy-consuming reactions in the cell like the synthesis of macromolecules, muscle contraction, regulatory networks, neuronal conduction or vesicle transport. Many essential processes in membranes critically depend on ATP as energy source as well, e.g. nutrient uptake in bacteria is often accomplished by ABC transporters and

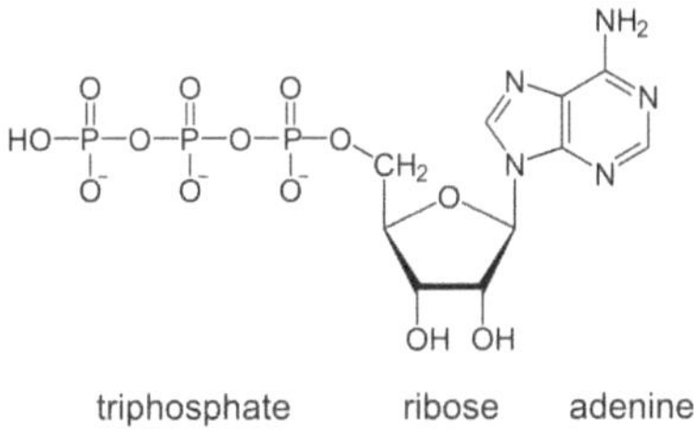

Figure 1.1: Structure of adenosine triphosphate.
The hydrolysis of the terminal phosphoric anhydride bond yields an energy of approximately 30 kJ/ mol. This energy is used in many biological processes to drive endergonic reactions.

1 General Introduction

eukaryotic cells use P-type ATPases to maintain their membrane potential.

The amounts of ATP required for life are impressive: humans have a daily turnover of ATP that equals their own body mass (Capaldi and Aggeler 2002).

To maintain a constant intracellular ATP concentration, cells have to recycle ATP permanently from ADP and inorganic phosphate.

There are several pathways cells can utilize for ATP synthesis. ATP can be generated by substrate level phosphorylation where the high phosphate transfer potential of metabolic intermediates e.g. phosphoenol pyruvate is used to attach a phosphate group to ADP. This way of ATP production is very fast, albeit not very efficient. Photophosphorylation employs the energy of light to create a proton-motive force ($\Delta\mu H^+$) across a lipid membrane which drives ATP synthesis by the F_1F_0 ATP synthase in the thylakoid membrane in chloroplasts. During oxidative phosphorylation, NADH or succinate are oxidized by the respiratory chain which concomitantly creates a proton-motive force. This force is then used by the F_1F_0 ATP synthase to produce ATP.

The sodium-motive force is used by some bacteria for ATP generation. These bacteria generate a Na^+ gradient by linking exergonic decarboxylation to endergonic Na^+ extrusion from the cytoplasm. In this so-called decarboxylation phosphorylation the Na^+ gradient is used by a Na^+-specific F_1F_0 ATP synthase for ATP production. In most cases ATP production via F_1F_0 ATP synthase delivers the majority of ATP in an organism.

The ATP synthase is not only a central hub in energy metabolism but it has a unique and fascinating rotational synthesis mechanism that has attracted researcher over the past four decades.

1.2 F-type ATPases

F-type ATPases or F_1F_0 ATP synthases are multi-subunit complexes found in most organisms. Their basic structure as well as the synthesis mechanism is conserved among all species. The subunit composition and nomenclature, however, varies between ATP synthases from different sources (see table 1.1). For ease of the reader subunit composition and names will always refer to the bacterial nomenclature. F-type ATPases are located in the cytoplasmic membrane of bacteria, in the inner membrane of mitochondria and in the thylakoid membrane of chloroplasts and cyanobacteria. Under most physiological conditions F-type ATPases synthesize ATP from ADP and P_i in response to an electrochemical gradient. The basic mechanism of energy conversion can be readily understood from the structure of this enzyme.

F-type ATPases consist of two structural parts, termed F_1 and F_0 which are held together by a peripheral and a central stalk. The subunit composition of the bacterial F_0 part is $ab_2c_{10\text{-}15}$

1.2 F-type ATPases

Table 1.1: Subunit composition of F_1F_0 ATP synthase from different sources.
A. The ATP synthase in mitochondria also contains the F_6, inhibitor, A6L, d-, e-, f-, and g-subunits, which have no equivalents in the enzymes from bacteria or chloroplasts. **B.** ATP synthases in *E. coli* and *Bacillus* PS 3 (both eight-subunit enzymes) have two identical copies of subunit b per complex. Purple non-sulfur bacteria and cyanobacteria appear to have nine different subunits, the extra subunits (known as b') being a homologue of b. Similarly, chloroplast enzymes are composed of nine non-identical subunits, and the chloroplast subunits known as I and II are the homologues of b and b' (Walker 1998).

Type	Bacteria	Chloroplasts	Mitochondria[A.]
F_1	α	α	α
	β	β	β
	γ	γ	γ
	δ	δ	OSCP
	ε	ε	δ
	-	-	ε
F_0	a	a	a or ATPase6
	b[B.]	b and b'	b
	c	c	c

with a total mass of roughly 150 kDa. All 3 subunits are integral membrane proteins. The c-subunits are assembled as a ring and the number of subunits can vary between different organisms from 10 - 15. The b_2-dimer and the a-subunit are laterally abutted to the c-ring. The F_0 part anchors the enzyme in the membrane and is mainly responsible for ion translocation and torque generation.

The F_1 part has a mass of approximately 350 kDa and is composed of subunits $\alpha_3\beta_3\gamma\epsilon\delta$. The α- and β-subunits are both able to bind ATP, however, only the binding site in the β-subunits is catalytically active. In contrast to the F_0 part all subunits as well as the entire F_1 complex are water-soluble. The F_1 part protrudes into the bacterial cytoplasm, the mitochondrial matrix or the chloroplast stroma where it catalyzes ATP hydrolysis or synthesis. A model for the F_1F_0 ATP synthase is shown in figure 1.2.

F-type ATPases have a remarkable and unique catalytic mechanism: They synthesize and hydrolyze ATP using the rotation of subunits which form a central "turbine wheel". An intriguing question in a rotational system is the definition of the rotor and the stator. Extensive cysteine cross-linking studies have defined the subunits moving as a complex in the F_1F_0 ATP synthase. It was convincingly shown by several groups that γε and the c-ring must rotate at the same time: γ and ε can be covalently interconnected without loss of ATP hydrolysis-driven proton pumping or ATP synthesis (Schulenberg et al. 1999; Tsunoda et al. 2001a). The groups of Fillingame and Cross conducted extensive cross-linking studies within the F_0 part, all of which are consistent with a rotation of the c-ring against the ab_2 subcomplex (Fillingame et al.

7

1 General Introduction

2000; Hutcheon et al. 2001).

In eukaryotic organisms and most bacteria the ATP synthase is required to produce rather than to hydrolyze ATP. In synthesis mode an electrochemical gradient drives ion translocation across the membrane. The ion flux generates a rotation of the c-ring that is transmitted to the catalytic sites in the F_1 part via the central stalk. The central stalk consists of the γ- and ϵ-subunits which are directly bound to the c-ring. The γ-subunit deeply protrudes in the hexagonal head piece formed by the $(\alpha\beta)_3$-subunits and ultimately leads to conformational changes in the catalytic nucleotide binding sites that result in the synthesis of ATP from ADP

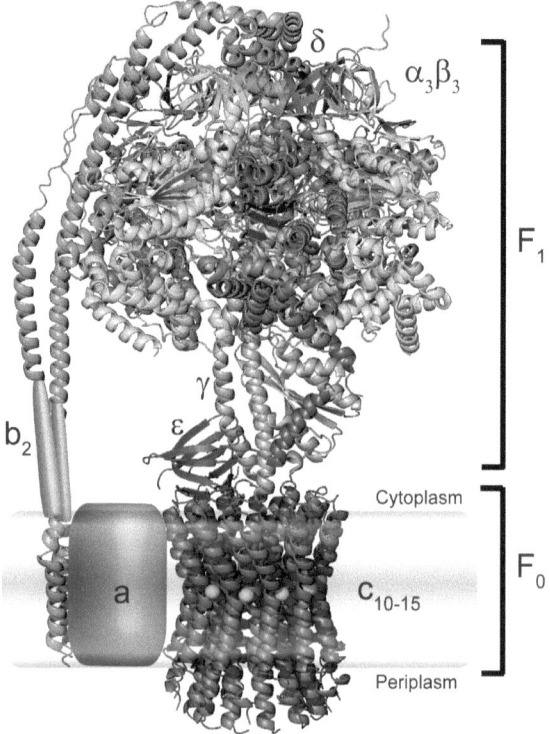

Figure 1.2: Subunit organisation and structure of the F_1F_0 ATP synthase.
The structures of individual subcomplexes are taken from the RCSB Protein Data Bank and assembled manually according to biochemical data. The structures used are from the c-ring of *I. tartaricus* (1CYE), the F_1 organisation of *E. coli* (1JNV), the δ-subunit of *E. coli* (2A7U), the peripheral stalk from bovine mitochondria (2CLY), and the membranous part of the b-subunit of *E. coli* (1B9U). No high-resolution structural data are available for the a-subunit and the hinge region of the b-subunit.

1.2 F-type ATPases

and P_i. This remarkable synthesis mechanism leads to a functional discrimination into a rotor, consisting of the γ- and ϵ-subunit as well as the c-ring and a stator with the subunit composition $\alpha_3\beta_3\delta b_2 a$. Under physiological conditions proper transmission of the rotation from c-ring to the F_1 head is essential to prevent energy loss. The fact that ions are only transported when ATP is synthesized is called coupling. As a result a constant coupling ratio between ATP generated and ions transported per revolution is achieved. Disturbing the interactions between stator and rotor *in vitro* e.g. by using detergents, can uncouple the enzyme. When coupling is lost, ions flow down the electrochemical gradient without leading to c-ring rotation and energy is dissipated.

Under certain conditions F-type ATPases can work as primary active ion pumps using the energy of ATP hydrolysis to transport ions across the membrane. This working mode is particularly important for facultative anaerobic bacteria that require the F-type ATPase to generate a membrane potential under anaerobic conditions when the respiratory enzymes are inactive (Dimroth and Cook 2004). Hence, both working modes of the F_1F_0 ATP synthase are physiologically relevant.

In the bacterial genome, the coding regions for the eight subunits are usually clustered in a single operon. The genes are arranged in the sequence uncIBEFHAGDC, which encode the i-protein, the F_0-subunits a, c and b and the F_1-subunits $\delta, \gamma, \alpha, \beta$ and ϵ, respectively (Fig. 1.3). The role of the i-gene is still not fully understood. It codes for a small membrane protein that co-purifies with the intact enzyme in substoichiometric amounts. It was recently shown that deletion of the uncI gene affects assembly of a stable c-ring in *P. modestum* suggesting a chaperone-like activity (Suzuki et al. 2007). Other reports have linked the i-protein with a function in Mg^{2+} uptake demonstrating channel like structures of the i-protein with the product of atpZ, a gene found in some bacterial unc operons (Hicks et al. 2003).

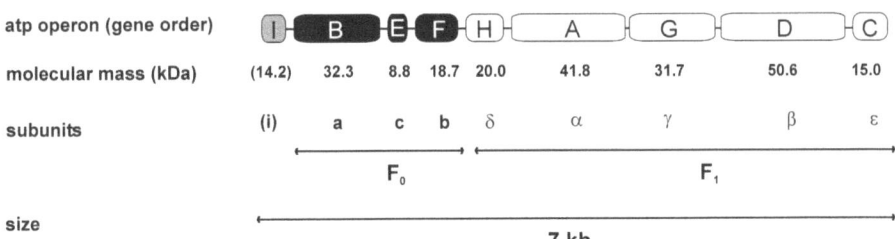

Figure 1.3: Organisation of the *Ilyobacter tartaricus* atp operon.
Genes encoding F_0-subunits are colored black and genes encoding F_1-subunits are depicted in white. The gene encoding the i-protein is drawn in light grey.

1.3 The F_1 part is responsible for ATP hydrolysis and synthesis

The F_1 part was first described as soluble factor which can be removed from beef heart mitochondria by low salt treatment. The newly discovered factor displayed ATPase activity and could restore ATP synthesis in membranes that had lost this ability (Penefsky et al. 1960; Pullman et al. 1960). Subsequently, a large body of data was collected on binding affinities, hydrolysis activities and ^{18}O-exchange experiments with the isolated F_1 part. It was demonstrated that the F_1 part displays strong negative cooperativity in substrate binding and concomitantly strong positive cooperativity in enzymatic activity. To explain these unusual properties, Boyer proposed the *binding change mechanism* in 1977 (Kayalar et al. 1977). The hypothesis proclaims that 3 catalytic sites sequentially undergo different conformations with corresponding different nucleotide affinities. The binding sites communicate with each other through the central position of the γ-subunit in a way that every subunit adopts one of the 3 possible conformations.

The synthesis cycle starts with the open conformation in which no nucleotide is bound. After binding of ADP and P_i the central stalk pivots 120° and the binding site changes its conformation to the loose state. A further turning of the central stalk results in the tight conformation in which ATP is formed from the bound ADP and P_i. The next step is the release of ATP leading to the open conformation and resetting the system into its starting position again (Boyer 1997).

Boyer's hypothesis is supported by first visualizations of the F_1F_0 ATP synthase by cryo-electron microscopy. The analysis showed that the α- and β-subunits alternate around a hexagon containing a central mass that was identified as the γ-subunit.

When the individual images were classified on the basis of their dominant features, they fell into three classes, with the γ-subunit being located at different $\alpha\beta$-pairs in each class (Gogol et al. 1990). In 1994 the high-resolution structure (2.8 Å) of a major part of the bovine heart F_1 part, including $\alpha_3\beta_3$ and part of the γ-subunit showed a 3-fold rotational symmetry of the $\alpha_3\beta_3$ head with the γ-subunit protruding deeply into the central cavity of the hexagonal headpiece. Detailed analysis of the single α- and β-subunits revealed three different conformations: one open, one partly open and the third one closed. The contacts between the C-terminus of γ and the N-termini of α and β form a hydrophobic sleeve allowing smooth rotation of the central stalk (Abrahams et al. 1994). All observed features in the structure of the F_1 part were in accordance with the *binding change mechanism*. Cross-linking of γ to the C-terminal domain of the β-subunit completely blocked activity (Aggeler et al. 1995). Cross and colleagues extended this work by demonstrating that, if the cross-link was subsequently

1.3 The F_1 part is responsible for ATP hydrolysis and synthesis

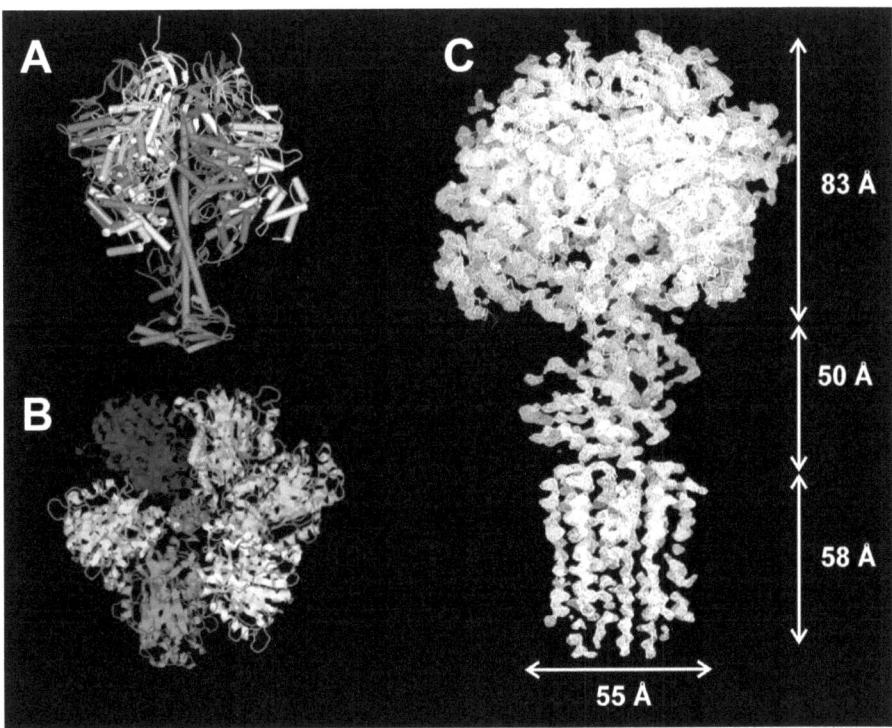

Figure 1.4: Structure of the F_1 ATPase from bovine mitochondria and yeast.
A. Side view of the F_1 ATPase from bovine mitochondria taken from Menz et al. (2001), demonstrating the central arrangement of the γ-subunit as a long helical structure with a β-barreled socket docking onto the F_0 part. **B.** The top view from the same enzyme is taken from Abrahams et al. (1994) and shows the arrangement of $\alpha_3\beta_3$ around the central γ-subunit. **C.** Electron density map of the F_1 c_{10} complex from *S. cerevisiae* at 3.9 Å, solved by X-ray crystallography (Stock et al. 1999). The structure gives an impression about the spatial arrangement in the intact enzyme. Subunit a, b and δ were lost during crystallization. Side chain assignment was impossible at this resolution.

released and reformed, γ became attached to different β-subunits (Duncan et al. 1995). In 1997 this predicted rotation was visualized for the first time by Noji et al. (1997). They attached a fluorescently labeled actin filament to the γ-subunit and observed the rotation of this filament by fluorescence microscopy. Under conditions where ATP binding was rate-limiting the rotation stalled every 120° as predicted by Boyer's *binding change mechanism*. The coupling of the mechanical rotation and the chemical reactions in the 3 binding sites are characterized today in great detail. The 120° step is dissected into an 80° substep, which is caused by ATP binding

1 General Introduction

in one site and ADP release in the other site (Yasuda et al. 2001). The dwell time following the large substep is due to ATP hydrolysis and P_i release from the same binding site. Finally, the P_i release triggers the 40° substep which completes the cycle (Adachi et al. 2007). Today,

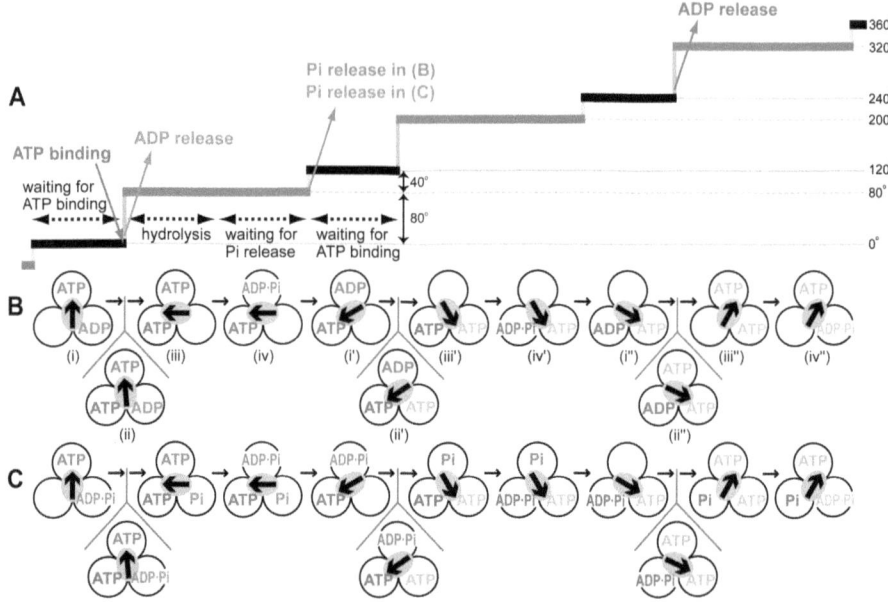

Figure 1.5: Coupling of rotation and catalysis in F_1 ATPase.
A. Schematic time course of stepping rotation. On the vertical axis is the rotary angle of γ is plotted, and time on the horizontal axis. **B.** Corresponding nucleotide states in the three catalytic sites. The three circles represent three β-subunits that each hosts a catalytic site. The central gray ellipsoid represents the γ-subunit, the thick arrow showing its orientation; the twelve o'clock position in (i) corresponds to 0° in A. Molecular species derived from the same ATP molecule depicted in the same tone. Small arrows indicate the progress in this major reaction pathway; the configurations (ii), (ii'), and (ii") shown below the major path represent the instant immediately after ATP binding, i.e., the start of a 80° substep. The time point of P_i release is not absolutely certain. The alternative pathway in which P_i release lags behind ADP release is displayed in **C**. Adapted from Adachi et al. (2007).

a large number of F_1 structures in different functional states is available (Abrahams et al. 1996 1994; Bowler et al. 2007; Braig et al. 2000; Cabezon et al. 2003; Kagawa et al. 2004; Menz et al. 2001). Combined with the data obtained from single molecule experiments the hydrolysis and synthesis cycle of the F_1 motor are described in atomic detail.

However, some questions concerning the impact of F_0 on F_1 still remain. One of these open questions is whether the central stalk acts as a spring during synthesis as the step size of the

1.3 The F_1 part is responsible for ATP hydrolysis and synthesis

c-ring and the F_1 head during one revolution are different. The c-ring performs a multistep rotation with the number of steps equalling the number of monomers in the ring, i.e. 10-15. The terminal parts of the γ-subunit which protrude deeply into the F_1 head on the other hand make a 3 step rotation. One suggestion was that the central stalk might acts as a spring accumulating torque produced by the c-ring and transmitting it to the $(\alpha\beta)_3$-subunits in a stepwise fashion. Junge and coworkers attached an actin filament to the γ-subunit and observed rotation with a constant angular speed when ATP was added to the isolated F_1 motor. Because the F_1 part was shown to cause rotation of the γ-subunit in discreet 120° steps under ATP limiting conditions, they concluded that the smooth rotation of the actin filament is due to a distortion of the central stalk that stores the torque produced in the $(\alpha\beta)_3$ head temporarily (Cherepanov and Junge 2001; Pänke et al. 2001). However, they observed rotation only in synthesis direction and molecular dynamics simulations suggest that a series of salt bridges is broken as the γ-subunit moves. As a result the restraints from F_1 towards the rotating central stalk are fairly constant (Bockmann and Grubmüller 2002).

Another open question is whether symmetry mismatch is required for function of the ATP synthase. None of the c-ring stoichiometries identified was a multiple of 3 and it was speculated that this might prevent the enzyme from slipping into a sort of dead center from which it cannot recover. However, Pogoryelov et al. (2005) isolated a c-ring with 15 subunits from the cyanobacterium *S. platensis* and disproved this hypothesis.

Much attention has been recently paid to the regulation of the F_1 motor. In contrast to the high conservation of the general features of F-type ATPases, the mechanisms for regulation of F_1 activity seems to be specifically adopted in different organisms.

When eukaryotic cells are deprived of oxygen, the electrochemical proton gradient across the inner mitochondrial membrane decreases and in order to maintain the gradient the ATP synthase starts to hydrolyze ATP. Under these conditions, the pH of the matrix drops due to acid production during substrate level phosphorylation.

In mitochondria, ATPase activity is regulated by the natural inhibitor protein IF_1, which binds to the ATP synthase in a pH-dependent manner. Bovine IF_1 can exist either as tetramer or as dimer, favoured by pH values above and below 6.5, respectively (Cabezon et al. 2000). The dimer is the active form of bovine IF_1. With its N-terminal inhibitory sequences it binds two F_1-parts, thereby blocking ATP synthesis (Cabezon et al. 2003). As the mitochondria re-energize in the presence of oxygen, the pH of the matrix increases, IF_1 dissociates and the enzyme can again synthesize ATP.

No homolog of IF_1 has been found in either chloroplasts or bacteria. ATPase activity of the chloroplast F_1F_0-ATP synthase is subject to complex regulation to prevent wasteful ATP hydrolysis in the dark. The enzyme is inhibited by Mg^{2+}-ADP (Digel et al. 1998), by the C-

1 General Introduction

terminus of the ε-subunit (Nowak and McCarty 2004; Richter et al. 1984), and by the oxidation/ reduction state of a disulfide bond within the γ-subunit (Nalin and McCarty 1984). Exposure of thylakoid membranes to light results in the generation of a $\Delta\mu H^+$ that stimulates the release of inhibitory bound ADP and relieves inhibition by the ε-subunit (Johnson and McCarty 2002). Reduction of the disulfide bond in γ further stimulates ATP synthesis (Nalin and McCarty 1984).

Like mitochondria, aerobic microorganisms are also faced with the challenge of blocking ATP hydrolysis activity when the cells become limited for oxygen or when $\Delta\mu H^+$ decreases. In bacteria, ATPase activity is regulated by a number of mechanisms involving inhibition by Mg^{2+}-ADP, $\Delta\mu H^+$ (Fischer et al. 2000; Zharova and Vinogradov 2003) and conformational changes in the ε-subunit (Kato et al. 1997; Kato-Yamada et al. 1999; Zharova and Vinogradov 2003).

The ε-subunit is divided into an N-terminal domain, which forms a flattened 10-stranded β-sandwich structure, and a C-terminal domain, which forms two α-helices running antiparallel to each other (Uhlin et al. 1997; Wilkens 1998). Its inhibitory effect is caused by the C-terminal domain as it undergoes a conformational rearrangement that involves transition of the two C-terminal helices between a hairpin downstate and an extended up-state (Tsunoda et al. 2001b). This transition has been verified through cross-linking studies (Suzuki et al. 2003). Without nucleotide, the ε-subunit is in the up-state and ATP induces the transition to the down-state, which is counteracted by ADP. The $\Delta\mu H^+$ stabilizes the up-state and in combination with ADP gears the enzyme for ATP synthesis (Kato-Yamada and Yoshida 2003).

The ε-subunit of some bacteria binds ATP and might act as a cellular sensor of ATP levels. The crystal structure of the ATP-bound ε-subunit from *Bacillus* PS3 shows that the two C-terminal α-helices fold into a hairpin structure residing on top of the N-terminal β-sandwich sheet. In the absence of ATP, the two α-helices extend to a more flexible structure. These data suggest that the ε-subunit can undergo an arm-like motion in response to changes in intracellular ATP concentration, and this may contribute to the regulation of ATP synthase *in vivo* (Yagi et al. 2007).

1.4 The F_0 part couples ion transport and torque generation

During ATP synthesis an electrochemical gradient is converted into torque by the F_0 part, whereas in the reverse direction ATP generated torque can be converted back to an electrochemical gradient.

There is much less known about the structure and the mechanism of the F_0 part than about the F_1 part. It is generally accepted that ion flux through the F_0 motor causes a rotation of the

1.4 The F_0 part couples ion transport and torque generation

c-ring. The main features of interconversion between chemical energy stored in a membrane potential and mechanical energy are understood: Ions flow through a poorly characterized channel in the a-subunit from the periplasm onto a binding site in the c-ring. Once the ion is bound the respective c-subunit leaves the a/c interface and makes almost an entire revolution until it enters the interface again from the other side. Here, the ion encounters the positive charge of a highly conserved arginine in the a-subunit which repels the bound ion. The binding site is then empty and the cycle can start again.

There are still many open questions concerning details of this mechanism e.g. how a membrane potential across the membrane is converted into a horizontal rotation and where the channels connecting the binding site on the c-ring to the periplasm and cytoplasm are located.

Unfortunately, structural information about the functionally most important subunit a is very scarce. Even the membrane topology is still under debate (Bjørbaek et al. 1990; Lewis et al. 1990; Long et al. 1998; Valiyaveetil and Fillingame 1998; Wada et al. 1999; Yamada et al. 1996). However, the overall architecture of the F_0 part is known from low resolutions electron micrographs: The oligomeric ring of c-subunits is flanked laterally by a single a-subunit (Mellwig and Bottcher, 2003). Subunit a is connected by the peripheral stalk to the $(\alpha\beta)_3$ head of F_1 part and the c-ring is linked at the cytoplasmic surface to the central stalk to form the rotor.

A high resolution structure of the c-ring from *Ilyobacter tartaricus* was presented in 2005 (Meier et al. 2005) and different parts of the b-subunit have been analyzed by X-ray crystallography and NMR (Dickson et al. 2006; Dmitriev et al. 1999).

These structures together with the large body of biochemical data describing driving forces for rotation and the ion path across the membrane point out the function of a small number of key amino acids which are indispensable for the mechanism of the F_0 motor.

1.4.1 Subunit a

Subunit a is the largest and mechanistically probably the most interesting polypeptide of the F_0 sector. Unfortunately, it is also the subunit with the least structural information and experimentally the most difficult to handle due to its high hydrophobicity.

The membrane topology is still under debate and models with 5 or 6 membrane spanning α-helices have been proposed (Jäger et al. 1998; Long et al. 1998; Valiyaveetil and Fillingame 1998; Yamada et al. 1996). There is general agreement about the topology for the two C-terminal helices. However, different topologies are suggested for the loop regions and the localization of the N-terminus. Subunit a does not seem to interact directly with the F_1 part, but with subunits b and c. Together with the c-ring, subunit a provides the ion translocation pathway (Deckers-Hebestreit and Altendorf 1996; Fillingame 1990; Howitt et al. 1996). Cross-linking

1 General Introduction

studies from Jiang and Fillingame (1998) showed a close interaction of the two C-terminal helices with the c-ring. The first of these two C-teminal helices (helix 4 according to Long et al. (1998)) contains the strictly conserved R210 (*E. coli* numbering). The residue R210 can not be replaced by any other amino acid. Even the conservative mutation R210K leads to complete loss of activity (Cain and Simoni 1989). Remarkably, the residues surrounding R210 are highly conserved as well. The consensus sequence $X_1RLX_2X_3N$ (X_1: small aliphatic (V, L, I) or F; X_2: T, A or F; X_3: A or G) is found in all organisms in the middle of helix 4. However, exchange of R210 with Q252 (i.e. the double mutant R210Q/ Q252R) resulted in active enzyme, albeit with a drastically reduced activity as compared to the wildtype (Hatch et al. 1995). There is general agreement that R210 ensures efficient repulsion of the coupling ion from the binding site on the c-ring into the cytoplasm (in synthesis direction). Mutagenesis has revealed E219 and H245 to be important as well, but not essential in the *E. coli* subunit a (Cain and Simoni 1986 1988 1989; Eya et al. 1991; Hartzog and Cain 1994; Howitt et al. 1990; Lightowlers et al. 1988).

Interestingly, alkaliphilic bacteria have a lysine at position 218 and a glycine at position 245. It was speculated that the histidine in *E. coli* and the lysine in *Bacillus* TA2 might act as proton catcher, each tuned for the optimal growth pH of its host. McMillan et al. (2007) showed that this hypothesis is at least partially correct. They could extend the activity profile of *Bacillus* TA2 ATP synthase to the neutral range by a K180H mutation in subunit a (corresponding to position 218 in *E. coli*). However, the authors pointed out that there is only a poor quantitative correlation between pK_a of the residue at position 180 and the pH optimum of the ATP synthase. The mode of interaction between subunit a and the oligomeric c-ring is one of the mysteries of the ATP synthase. The interface has to provide enough interactions to ensure the integrity and coupling of the enzyme on one hand, on the other hand the friction during rotation (30-40 Hz) must be low to achieve the extraordinarily high efficiency (near 100 %) of this splendid molecular machine.

1.4.2 Subunit b

Two identical b-subunits wind up as a coiled coil and form the peripheral stalk. Dimerization of the b-subunits is thought to be an early event necessary for enzyme assembly and function (Sorgen et al. 1998ab). The two b-subunits exist in an extended α-helical conformation, spanning from the periplasmic side of the membrane to near the top of F_1 where they interact with the δ-subunit. The b-subunit is divided in several sections, named by their main tasks: The transmembrane domain consists of residues 1-24 (*E. coli* numbering) and forms an α-helix that anchors the subunit in the membrane. Its structure has been solved by NMR in a mixed polar solvent (Dmitriev et al. 1999). Residues 25-52 form the tether domain which is the least understood part of the b-subunit. It is known to interact with the a-subunit and to play a role in coupling F_1 with F_0 rotation. The dimerization domain, reaching from residue 53 to 122, forms a right-handed coiled-coil that clamps the two b-subunits together. Finally, the C-terminal δ-binding domain links the b-subunits to the F_1 part.

High resolution structures are available for the isolated transmembrane domain and for the isolated dimerization domain from the *E. coli* enzyme. Recently, the soluble part from the bovine ATP synthase containing subunits OSCP, b, d and F_6 has been solved and it shows an extended curving shape (Dickson et al. 2006). Various features suggest that the peripheral stalk is rigid rather than resembling a flexible rope. This rigidity is consistent with the proposed function of the b-subunit to counteract the rotation of the rotor and keep the $(\alpha\beta)_3$ hexamer in place. The role of subunit b as elastic energy storage has been raised and different models are discussed (Blum et al. 2001). Recent work by Diez et al. (2004) suggests that the interaction between b_2 and F_1 is not strong enough to withstand the elastic strain required for the synthesis of ATP in one single step. Although, subunit b has been demonstrated to be indispensable for ion transport (Schneider and Altendorf 1985), no direct participation was found. In a recent work it was shown that the helical part of subunit b is sufficient to drive coupled ion transport across the membrane, although not at full speed (Greie et al. 2004).

1.4.3 Subunit c

Subunit c of *E. coli* consists of only 79 amino acids and is highly hydrophobic. Due to its extreme hydrophobicity it can be readily extracted with organic solvents and was therefore termed proteolipid in earlier reports. It plays a key role in the ion translocation mechanism by shuttling the ions between the periplasmic and the cytoplasmic reservoir. The protein folds like a hairpin with two transmembrane-spanning α-helices separated by a conserved polar loop region, which is exposed to the cytoplasm. A highly conserved carboxyl group (D61 in *E. coli*) in the middle of the membrane is an essential part of the binding site for the coupling ions. The subunits assemble as a ring consisting of 10-15 copies, depending on the organism. The solution structure of the *E. coli* c-monomer was determined using NMR in a chloroform/ methanol/ water (4:4:1) mixture at pH 8.0 and 5.0 (Rastogi and Girvin 1999ab). At these pH values the protein folds into two extended, slightly bent α-helices, which are connected by a polar loop. The two structures adopt different conformations at low or high pH, the major distinction being a 140° turn of helix II with respect to helix I. It is assumed that this conformational change is elicited by the protonation or deprotonation of D61. Based on these results a mechanistic model was proposed describing how proton translocation across F_0 could be coupled to the

1 General Introduction

generation of torque (Rastogi and Girvin 1999b). In this model, torque generation is assumed to be driven by the clockwise rotation of helix II of subunit c (as viewed from the cytoplasm) when D61 becomes protonated. However, the conformational change accompanying the pH change was not observed in the c-subunit of *Bacillus* PS3. In addition the helix packing seen in the c-ring structure from *I. tartaricus* and the K-ring of *E. hirae* is so compact that such a large rotation would most likely disrupt the entire ring. As a result this hypothesis is regarded obsolete today.

The Na^+ binding site

The most valuable insights into function and structure of the c-ring have come from the Na^+-dependent ATP synthases from *P. modestum* and *I. tartaricus*. Already early genetic experiments by Kaim et al. (1997) showed that in addition to the conserved E65 and S66 and Q32 are essential residues for Na^+ binding in *P. modestum*. Today, the binding site in the Na^+-dependent F-type ATPase from the closely related *I. tartaricus* is known in atomic detail (Fig. 1.6). The conserved carboxyl group is delivered by E65 and is part of a complex hydrogen bonding network, which presumably stabilizes the accurate orientation of the other side chains participating in Na^+ binding. The Na^+ ion bridges two adjacent c-monomers, stabilizing the strong interactions between the c-ring monomers of Na^+-dependent enzymes. How well the binding site is tailored for Na^+ ions is observed when replaced by Li^+ ions: The stability of the ring drastically decreases (Meier and Dimroth 2002).

The Na^+ binding sites from the c-ring of *I. tartaricus* and the K-ring from the Na^+-depended V-type ATPase of *Enterococcus hirae* are shown in figure 1.6. Both structures share an amazing homology: With exception of the additional Q65 in *E. hirae* the binding site is essentially the same in both organisms (Murata et al. 2005). The Na^+ ion in the structure from *I. tartaricus* has four ligands: E65, Q32, S66 and the backbone carbonyl of V63. In the structure of the K-ring from *E. hirae* a second glutamine (Q65) provides a fifth coordination site for the Na^+ ion. This locked conformation mimics most likely the situation outside the a/ c interface, when the subunits face the lipidic environment and the Na^+ is tightly bound. To allow release or binding of the ion in the a/ c interface the conformation of the binding site very likely changes significantly. One of the factors responsible for this change is almost certainly the conserved aR226 (aR210 in *E. coli*) in the a-subunit.

The properties of the binding site of *P. modestum* were intensively investigated using dicyclohexylcarbodiimide (DCCD). DCCD reacts with the protonated carboxylic acid in the binding site. This allows titration of E65 in the absence of Na^+. The rate of DCCD labeling decreases with increasing pH, displaying a sigmoidal curve with an inflection point at pH 6.8, correspond-

1.4 The F_0 part couples ion transport and torque generation

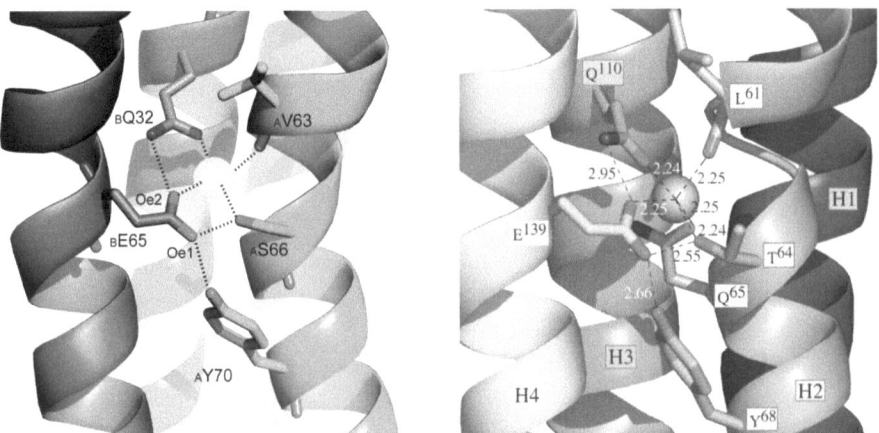

Figure 1.6: Comparison of the Na$^+$ binding sites from the c-ring of *I. tartaricus* and the K-ring of *E. hirae*.
On the left side the binding pocket from the *I. tartaricus* c-ring is shown and on the right the binding pocket from the *E. hirae* K-ring. The spatial arrangement of the Na$^+$ binding sites is very similar. In both organisms a complex hydrogen bonding network stabilises the conformations of the side chains in the binding pocket. The K-ring has an additional residue, Q65, involved in Na$^+$ coordination.

ing to the pK$_a$ of E65. If Na$^+$ is added to the solution, the dependency is shifted towards the acidic range (Kluge and Dimroth 1993a). This reflects a competition of the two ions for the binding site. When the experiment is repeated with the *E. coli* enzyme a broad peak with high labeling is observed between pH 7.5 and 10. This behavior of the H$^+$-dependent enzyme can not be explained by group protonation of E61. One suggestion already expressed by Boyer in 1977 implies that a hydronium ion (H$_3$O$^+$) might be bound in the binding site resulting in such an unusual labeling profile. Biochemical data clearly show that the binding sites in Na$^+$ dependent enzymes and H$^+$-dependent enzymes are different (von Ballmoos and Dimroth 2007). However, it will be difficult to obtain direct evidence for the existence of a hydronium ion without a high resolution structure.

The structure of the *I. tartaricus* c-ring

The X-ray structure of the c$_{11}$-ring from *I. tartaricus* was successfully solved to a resolution of 2.4 Å by Meier et al. (2005). The structure confirmed many features observed earlier in the moderate resolution structures obtained by AFM and electron cryomicroscopy. The high resolution however, allows assignment of the amino acid side chains which created a new order

1 General Introduction

of accuracy.

The electron density map of a single c-ring shows a cylindrical, hourglass-shaped protein complex of ~ 70 Å in height, and with an outer diameter of ~ 40 Å in the middle and ~ 50 Å at the top and bottom. Eleven c-subunits, each composed of two membrane-spanning α-helices forming a hairpin, are arranged around an 11-fold axis, creating a tightly packed inner ring with the N-terminal helices (Meier et al. 2005). The dense packing is reflected in the sequence by the conserved GxGxGxG pattern found in bacterial, chloroplast and mitochondrial c-subunits (Vonck et al. 2002). The C-terminal helices pack into the grooves formed between N-terminal helices, producing an outer ring. The N- and C-terminal helices are connected by a loop formed by the highly conserved sequence R45, Q46 and P47, which is exposed to the cytoplasmic surface. The C-terminal helices are shorter than the N-terminal helices and show an unexpected break after A81. The break is followed by the residues N82 and P83 which do not form an α-helix. From F84 until the C-terminus an α-helix is formed again, however with a tilt of approximately 40° and offset 10 Å relative to the preceding helix. The cause for this unusual arrangement might lie in the fixation of F84 by a hydrophobic clamp consisting of residues V10 and P83 of the same c-subunit and L11, L79 and A15 of the neighboring subunit. Moreover, N82 is fixed by hydrogen bonds between S14 and the amide group of F84. As a result of the distortion at the C-terminus, the position of the outer helix might be slightly altered for an optimized binding site. If the C-terminus of *I. tartaricus* is truncated at F84, the affinity for Na^+ is reduced one order of magnitude (personal communication Ch. von Ballmoos).

Interestingly, the number of c-subunits forming the ring is not fixed but varies among species. In a complete revolution of the rotor each c-subunits transports one ion across the membrane and concomitantly three molecules of ATP are formed. Hence the H^+ (Na^+)/ATP ratio varies between 3.3 (c_{10}) and 4.7 (c_{14}). The physiological relevance of the different ratios and the obvious prerequisite of a mismatch are topics of actual discussions.

1.5 Na^+-dependent F_1F_0 ATP synthases

1.5.1 *Propionigenium modestum* and *Ilyobacter tartaricus*

In the early 1980's, Schink and coworkers isolated and characterized marine bacteria from anoxic brackwater in Venice (Schink, 1984; Schink and Pfennig, 1982).

One of the organisms they isolated was a rod-shaped, strictly anaerobic, non-motile Gram-negative bacterium, which grows on the fermentation of succinate to propionate and CO_2 (Hilpert et al. 1984). The only exergonic reaction in the energy metabolism of this so-called *P. modestum* is the decarboxylation of (S)-methylmalonyl-CoA to propionyl-CoA and CO_2.

During this decarboxylation reaction Na$^+$ ions are translocated across the membrane and generate an electrochemical Na$^+$ gradient. This Na$^+$ gradient was found to be the only possible energy source for ATP generation and consequently a Na$^+$-dependent F-type ATP synthase was isolated and characterized shortly afterwards (Laubinger and Dimroth 1987 1988).

The close relative *I. tartaricus* has similar morphological and physiological properties as *P. modestum*. Both F-type ATP synthases are extremely similar but because of the higher cell yields achieved with *I. tartaricus* it was used for many experiments instead of *P. modestum* (Neumann et al. 1998).

General features of the *P. modestum* ATP synthase

The *P. modestum* F_1F_0 ATP synthase is homologous to other F-type ATPases (Dimroth 1997). The eight different subunits are organized in a single operon on the chromosome.

Compared to *E. coli*, the sequence conservation within subunits is rather low; exceptions are the nucleotide binding subunits α and β, which display a high degree of homology across all species.

The most striking feature is the selectivity of the enzyme for Na$^+$ ions. In the absence of Na$^+$ ions no hydrolysis activity can be measured. Only in a narrow pH range around 6.5, the enzyme can alternatively pump protons across the membrane. Under more physiological condition, the enzyme shows a strong dependency on Na$^+$ ion concentration and exhibits a functional pH range from 7 to 9. Li$^+$ can act as an alternative coupling ion, however with a K_m which is about ten times higher than for Na$^+$ (Kluge and Dimroth 1993ab).

As expected, the ion specificity is solely determined by the F_0 part. This was demonstrated by the construction of *E. coli* (F_1 part) and *P. modestum* (F_0 part) hybrid enzymes displaying ATP hydrolysis-dependent Na$^+$-transport (Gerike et al. 1995; Kaim and Dimroth 1993 1994).

1.6 The ion pathway through the membrane

Both, the ATP synthesis and hydrolysis are intimately coupled to ion transport across the membrane, which is accomplished by the membrane-embedded F_0 part. There is general agreement that during ATP synthesis the coupling ions enter the membrane from the periplasm through a channel in subunit a, from where they are dislodged to the binding site on subunit c, which is located in the middle of the membrane. The ions remain on the binding site, until they are released towards the cytoplasm. The location of the release channel is still unclear. Experiments with Na$^+$-dependent enzymes suggest a location within the c-ring (Meier et al. 2003; von Ballmoos et al. 2002ab). However, in the c-ring structure a channel could not be identi-

1 General Introduction

fied. Alternatively, a second half channel within the a-subunit is proposed for H^+-translocating enzymes. Whereas clear evidence for the existence of the periplasmic channel of subunit a is available, the experimental data for a cytoplasmic half channel within subunit a are rather poor (Angevine et al. 2003; Kaim et al. 1998). It is possible that both subunit a as well as subunit c contribute to the channel and that it only forms in the a/ c interface. To address this question a structure of the a/ c interface will be required.

1.7 Driving forces for F-type ATPases

The chemiosmotic theory was formulated by Peter Mitchell in 1966 (Mitchell 1979). It explained the coupling between respiratory chain and ATP synthase by circulation of protons and was based on 4 main postulates: (1) The ATP synthase is a reversible proton-motive ATPase of characteristic H^+/ P_i stoichiometry. (2) Respiratory and photoredox chains are proton-motive systems of characteristic H^+/ $2e^-$ stoichiometry, and appropriate polarity. (3) There are proton-linked solute porter systems for osmotic stabilisation and metabolite transport. (4) Systems 1 to 3 are plugged through a topologically-closed membrane of low permeability to solutes in general and to H^+ and OH^- ions in particular.

In the same year as the chemiosmotic theory was postulated, Jagendorf and Uribe (1966) demonstrated that ATP synthesis could be driven by ΔpH. In the following years an impressive number of reports was published and they all supported Mitchell's hypothesis of the proton-motive force ($\Delta\mu H^+$) fueling ATP synthesis.

The $\Delta\mu H^+$ is composed of two components which are thermodynamically equivalent: the membrane potential ($\Delta\psi$) and the concentration difference of protons (ΔpH).

$$\Delta\mu H^+ = \frac{k_B \cdot T}{q} \times ln\left(\frac{c_1}{c_2}\right) + \Delta pH$$

Implicit in this equation is the assumption that membrane potential and transmembrane ion gradients are equivalent driving forces for ATP synthesis. It is however experimentally not easy to separate ΔpH and $\Delta\psi$ accurately in an experimental setup. It turned out that the acid bath procedure developed by Jagendorf and Uribe to create solely a ΔpH, created a $\Delta\psi$ concomitantly (Kaim and Dimroth 1999).

In this procedure thylakoid membranes are equilibrated with succinate buffer pH 5 and rapidly diluted into a buffer at pH 8. Kaim and Dimroth (1999) showed that the predominant succinate species at pH 5, the monoanion, is membrane-permeable and quickly equilibrates with the thylakoid membranes. At pH 8 the predominant species is the dianion which is membrane-impermeable. If the monoanion diffuses along its concentration gradient into the external buffer, it will be quickly deprotonated and removed from equilibrium for monoanions. Hence, a constant electrogenic flow of succinate monoanions out of the membranes occurs which creates a $\Delta\psi$ of up to -140 mV.

Kaim and Dimroth (1999) demonstrated that the $\Delta\psi$ is the principal driving force for F-type ATP synthases from *E. coli*, spinach chloroplasts and *P. modestum*. A concentration gradient (ΔpH or ΔpNa, respectively) hardly has an impact on synthesis rates. Some of these results could not be confirmed by this PhD thesis and will be discussed in chapter 3 on page 49.

The notion that $\Delta\psi$ is mainly responsible for ATP synthesis came from studies with the isolated F_0 motor from *P. modestum*. Sodium ion uptake into F_0-containing liposomes was not observed at $\Delta\psi$ values < -40 mV. The Na^+ ion transport rate increases exponentially with increasing membrane potentials approaching saturation at $\Delta\psi$ > -120 mV (Kluge and Dimroth 1992). Upon reversal of the membrane potential, the direction of Na^+-transport changes to the export of Na^+ from the F_0 liposomes.

In the absence of a membrane potential, the F_0 motor is in an idling mode performing oscillations of the rotor versus the stator in either direction, which is characterized by Na^+ ion exchange across the membrane. Upon applying voltage (i.e. $\Delta\psi$), the rotation is rectified and Na^+ ions are transported unidirectionally across the membrane. The presence of a large Na^+ concentration gradient does not initiate Na^+ uptake as observed by Kluge and Dimroth (1992). Surprisingly, even a ΔpNa opposing a $\Delta\psi$ of the same magnitude had no effect on initial uptake rates. However, when the direction of Na^+-transport was reversed and efflux was measured, either $\Delta\psi$ or Δ pNa were able to elicit transport.

These findings propose different behavior of the ATP synthase from *P. modestum* in synthesis and hydrolysis direction and will be presented and discussed in detail in chapter 3 of this thesis.

In alkaliphilic bacteria ΔpH and $\Delta\psi$ are oriented in opposite directions, due to the high extracellular pH. These bacteria are faced with the challenge of synthesizing ATP at low $\Delta\mu H^+$. Krulwich and co-workers have proposed that a proton pathway generated by respiration-dependent proton pumping and the ATP synthase may in fact be localized on or in the cytoplasmic membrane (Krulwich 1995). Based on this model, discrepancies between the bulk $\Delta\mu H^+$ and the rate of steady state oxidative phosphorylation would not need to be taken into account. The observations made by Krulwich and others (Branden et al. 2006; Michel and Oesterhelt 1980b) would slightly modify the chemiosmotic hypothesis: Instead of the bulk pH, the actual proton concentration at the lipid membrane is crucial for ATP synthesis. Even though the concept of localized proton transfer reactions along membrane surfaces was proposed long ago, it has yet to enter the awareness of researchers.

1 General Introduction

1.8 Model of the F_0 motor

Based on experimental data available for the *P. modestum* enzyme Xing et al. (2004) presented a model for the mechanism of ion translocation through the F_0 part. They quantitatively described a mechanochemical model for F_0 rotation with the corresponding energies. Key features of their model are a horizontal component of the membrane potential which lowers the energy of the Coulomb interaction between the cE65 in the empty binding site and the aR226 in the a-subunit. Another critical step in their model is the hydration of the empty binding site as it enters the periplasmic channel: they suggest that the hydration energy delivers part of the energy required to break the cE65/ aR226 interaction. In addition this energy prevents back rotation of the c-ring, due to the high energy required for dehydration. The sequence of events occuring in a binding site when it rotates along the a/ c interface in synthesis direction is shown in figure 1.7. The crucial steps during ion translocation in ATP synthesis direction in the a/ c interface can be divided into four different zones. For H^+ ATPases the events in each zone are as follows: (1) The occupied rotor site enters the interface and releases the proton through the outlet channel into the cytoplasm with high pH. The deprotonation of the binding site prevents backwards rotation into the lipid phase and acts as a molecular ratchet. (2) The negative charge of the binding site is compensated by the stator arginine. (3) The binding site enters the periplasmic channel and binds a proton due to the higher proton concentration as

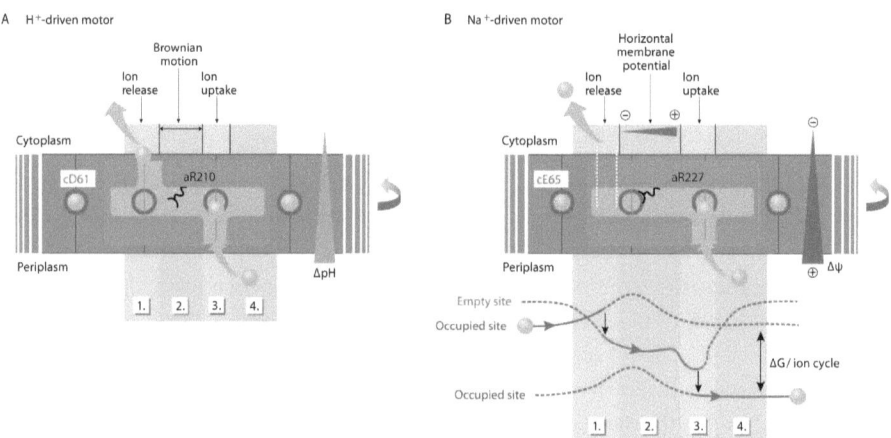

Figure 1.7: Model for torque generation in the H^+- and Na^+-translocating F_0 motor.
A. Two-channel model with a ratchet-type mechanism for H^+-dependent enzymes. **B.** Push-and-pull model for Na^+-dependent enzymes (Dimroth et al. 2006).

compared to the outlet channel. (4) The loaded binding site moves out of the interface into the lipid bilayer, whereby the next binding site enters the interface and follows the events described in (1). For Na$^+$ ATPases a mathematical model was developed and free energy (ΔG) values calculated for an occupied and free binding site respectively, which is depicted in the lower part of figure 1.7 B.

The upper part shows the events taking place in the a/ c-subunit interface. Arrows indicate where a Na$^+$ ion is released or taken up, respectively. (1) In ATP synthesis direction, an occupied rotor site enters the interface from the left and releases its bound Na$^+$ ion towards the cytoplasm. This process is aided by the stator arginine. (2) The stator arginine compensates for the empty binding site which is now negatively charged. Xing et al. (2004) proposed a horizontal component of the membrane potential pulling the arginine to the left and pushing the glutamate to the right. Therefore, the electrical component of the ion-motive force would determine the direction of rotation from left to right. (3) The hydration of the binding site within the inlet channel stabilizes this conformation and allows loading of the binding site from the periplasm. Movement of the binding site from zone 2 to 3 pulls the next rotor site into the a/ c interface as described in (1). (4) The binding site that has been occupied from the periplasm is allowed to rotate out of the interface into the lipid phase. This step is promoted by a push mechanism during the events described in (2).

1.9 Aim of this work

The central role of the ATP synthase in the metabolism has attracted scientists over the last decades to explore its structure and mechanism. Today, it is generally accepted that ATP synthesis is accomplished by a series of energy conversions. The electrochemical membrane gradient is converted into rotation of the c-ring by the F_0 part. The resulting torque is subsequently transmitted to the F_1 part and transformed into chemical energy in the nucleotide binding sites.

A large amount of kinetic as well as structural data has been collected that gives insight into the mechanism of these energy conversions. However, the mechanism of torque generation in the F_0 part is still enigmatic.

In this thesis a method is described that allows accurate kinetic investigation on the F_0 part from H$^+$ translocating ATP synthases (chapter 2 on page 27).

Investigation of the driving forces required for ATP synthesis in the Na$^+$-dependent enzyme from *P. modestum* lead to the discovery that ΔpNa is essential. Additionally it was shown that during ATP synthesis the enzyme displayed an apparent K_D for Na$^+$ of 35 mM which is 70 x higher than during ATP hydrolysis. We analysed the requirements for the H$^+$-dependent

1 General Introduction

enzyme from *E. coli* and observed a similar mismatch between ATP synthesis and hydrolysis (chapter 3 on page 49).

Further examination of the F_0 part revealed fundamental differences between ion transport in hydrolysis or synthesis direction through the F_0 part which are described in chapter 4 on page 67.

2 Δψ and ΔpH are equivalent driving forces for proton transport through the isolated F_0 complexes of ATP synthases

2.1 Abstract

The membrane-embedded F_0 part of ATP synthases is responsible for ion translocation during ATP synthesis and hydrolysis. Here, we describe an *in vitro* system for measuring proton fluxes through F_0 complexes by fluorescence changes of the entrapped fluorophore pyranine. Starting from purified enzyme, the F_0 part was incorporated unidirectionally into phospholipid vesicles. This allowed analysis of proton transport in either synthesis or hydrolysis direction with Δψ or ΔpH as driving forces. The system displays a high signal-to-noise ratio and can be accurately quantified. In contrast to ATP synthesis in the *E. coli* F_1F_0 holoenzyme, no significant difference was observed in the efficiency of ΔpH or Δψ as driving forces for H^+-transport through F_0. Transport rates showed linear dependency on the driving force. Proton transport in hydrolysis direction was about 2350 $H^+/$ (s × F_0) at Δψ of 120 mV, which is approximately twice as fast as in synthesis direction. The chloroplast enzyme was faster and catalyzed H^+-transport at initial rates of 6300 $H^+/$ (s × F_0) under similar conditions.

The new method is an ideal tool for detailed kinetic investigations of the ion transport mechanism of ATP synthases from various organisms.

2.2 Introduction

F_1F_0 ATP synthases are complex molecular machines that reside in the inner membrane of mitochondria, the thylakoid membrane of chloroplasts or the cytoplasmic membrane of bacteria. They catalyze the synthesis of ATP from ADP and P_i with the energy stored in an

2 $\Delta\psi$ and ΔpH are equivalent driving forces for the F_0 part

electrochemical ion gradient across the membrane (Boyer 1997).

In fermenting bacteria where respiratory enzymes are not active, ATP synthases can work in reverse as ATPases to generate the membrane potential ($\Delta\psi$) required for cell viability. Hence, both working modes of F_1F_0 ATP synthases are physiologically relevant. The construction of the F_1F_0 ATP synthases is bipartite: The F_1 part with the subunits $\alpha_3\beta_3\gamma\epsilon\delta$ harbors the catalytic sites for ATP synthesis and hydrolysis and protrudes into the aqueous compartment of the cell. The F_0 part in its simplest bacterial form consists of subunits ab_2c_{10-15}. It is membrane-embedded and catalyzes ion transport across the membrane. The number of c-monomers in the oligomeric c-ring varies between species and concomitantly affects the ion/ ATP ratio during ATP synthesis or hydrolysis (Meier et al. 2007). ATP synthesis measurements with the purified enzymes of *E. coli* and spinach chloroplasts have shown that electric potential and proton gradient are not equivalent driving forces (Fischer and Gräber 1999). It is unknown, however, if these discrepancies are F_0 intrinsic properties or if they are related to the holoenzyme.

To understand the principle of proton pumping in either direction and to investigate asymmetries between the synthesis and hydrolysis mode of F_1F_0 ATP synthases, a defined system is needed, in which unidirectional proton transport can be measured. In this work we developed a method to measure unidirectional proton transport through the F_0 part in synthesis and hydrolysis direction, energized selectively by either a proton gradient (ΔpH) or an electrical potential ($\Delta\psi$). The assay is based on the hydrophilic fluorophore pyranine which is entrapped inside proteoliposomes and allows proton transport to be quantified by monitoring fluorescence changes. The relative simplicity of the system accounts for a high reproducibility and straightforward determination of all parameters required for a quantification of H^+-transport rates. It is moreover generally applicable to ATP synthases from various origins and displays several advantages over described procedures.

2.3 Material and Methods

Antibody against subunit b was a generous gift of Karl-Heinz Altendorf, University of Osnabrück, Germany. Chemicals were purchased from Fluka (Buchs, Switzerland) if not otherwise indicated.

Enzyme purification and reconstitution into liposomes. For expression of the *E. coli* F_1F_0 ATPase, plasmid pBWU13 encoding the entire atp operon and an N-terminal His-tag on the β-subunit was used (Moriyama et al. 1991). The plasmid was transformed and expressed in *E. coli* strain DK8 which lacks the entire ATP operon (ΔuncBEFHAGDC) (Klionsky et al. 1984). Cells were grown overnight at 30 °C in LB medium supplemented with 100 µg/ l ampicillin and 20 µg/ l tetracyclin.

F_1F_0 ATPase was isolated and purified following a slightly modified procedure of Ishmukhametov et al. (2005). Briefly, 5 g cells were resuspended in 20 ml lysis buffer (200 mM Tris-Cl pH 7.8, 100 mM KCl, 5 mM $MgCl_2$, 0.1 mM K_2-EDTA and 2.5 % glycerol) and subsequently passed twice through a French press cell at 100 MPa. Unbroken cells were removed by centrifugation for 5 min at 8,000 g. Membranes were collected by centrifugation of the supernatant for 30 min at 200,000 g. For solubilization, membranes were suspended in 10 ml extraction buffer (50 mM Tris-Cl pH 7.5, 100 mM KCl, 250 mM sucrose, 40 mM ϵ-aminocaproic acid, 15 mM p-aminobenzamidine, 5 mM $MgCl_2$, 0.1 mM K_2-EDTA, 0.2 mM DTT, 0.8 % phosphatidylcholine, 1.5 % octylglucopyranoside, 0.5 % sodium deoxycholate, 0.5 % sodium cholate, 2.5 % glycerol, 30 mM imidazole) and gently stirred for 1 h at 4 °C. Unsolubilized material was removed by centrifugation for 45 min at 200,000 g and the supernatant passed through a 0.22 µm filter (Millipore AG, Zug, Switzerland). The filtrate was loaded on a Ni-IDA column (GE Healthcare, Glattbrugg, Switzerland) and washed with 10 column volumes of extraction buffer. The F_1F_0 ATPase was eluted with extraction buffer containing 400 mM imidazole. The elution fractions were analyzed on SDS-PAGE and stored in liquid nitrogen.

Chloroplast F_1F_0 ATP synthase was isolated from spinach leaves as described (Turina et al. 2003). Briefly, 1.5 to 2 kg of spinach leaves were homogenized in a sucrose containing buffer with a blender. Subsequent filtration and centrifugation steps were performed to obtain a thylakoid membrane preparation (chlorophyll concentration of 5 mg/ ml) and the samples were stored at -20 °C. The ATP synthase was solubilized from the membranes, fractionated by ammonium sulphate precipitation and subsequent sucrose density gradient centrifugation as described (Turina et al. 2003). The purified protein samples (5 mg protein/ ml) were stored in liquid nitrogen.

Purification of *E. coli* F_1. *E. coli* F_1 complex was purified as described previously (Senior et al. 1979).

Reconstitution of F_1F_0 ATP synthase from *E. coli* and chloroplasts. The F_1F_0 ATPase from *E. coli* and chloroplasts was reconstituted following a slightly modified procedure described in Ishmukhametov et al. (2005). Soybean phosphatidylcholine (Sigma-Aldrich, Buchs, Switzerland) was dissolved at a concentration of 30 mg/ ml in buffer A (10 mM Tricine-NaOH pH 8.0, 2.5 mM $MgCl_2$, 0.1 mM Na_2-EDTA, 0.2 mM DTT) and sonicated at 7.5 µ for 2 x 30 s on ice using a tip sonicator (Sanyo MSE Soniprep, München, Germany) to form unilamellar liposomes. The suspension was adjusted to 1 % sodium cholate from a 10 % stock solution and mixed with F_1F_0 ATPase (lipid : protein ratio (w/ w) 1:50). The mixture was incubated

2 $\Delta\psi$ and ΔpH are equivalent driving forces for the F_0 part

for 20 min on ice and subsequently 1 ml was loaded on a PD-10 column (GE Healthcare, Glattbrugg, Switzerland), preequilibrated with buffer A. Turbid fractions were pooled and the proteoliposomes collected by centrifugation for 45 min at 200,000 g at 4 °C.

Preparation of F_0 liposomes. F_1F_0 proteoliposomes prepared as described above were dialyzed overnight against 1000 volumes of stripping buffer (0.5 mM Tricine pH 8.5, 0.5 mM Na_2-EDTA) at 4 °C to remove the F_1 part. The sample was diluted with stripping buffer and F_0 liposomes were collected by centrifugation for 1 h at 200,000 g at 4 °C. The liposomes were resuspended at 60 mg/ml in buffer B (2 mM MOPS-NaOH pH 7.2, 2.5 mM $MgCl_2$, 50 mM Na_2SO_4 or 50 mM K_2SO_4) including 1 mM pyranine and frozen in liquid nitrogen for 10 min, thawed in cold water and sonicated twice for 10 s in a water bath sonicator. The freeze/thaw/sonication procedure was repeated once. To remove external pyranine, the liposomes were loaded on a PD-10 column equilibrated with buffer B containing either Na_2SO_4 or K_2SO_4. The turbid yellowish fraction was collected and concentrated by centrifugation (30 min at 200,000 g at 4 °C). The liposomes were resuspended at a lipid concentration of 120 mg/ml in the same buffer and stored at 4 °C. No loss in activity was observed within 7 days. Depending on the inner salt, they were denoted Na^+- or K^+-liposomes. For equilibration at the desired pH, liposomes were diluted 20-fold in the respective buffer and incubated for 16-72 h at 4 °C.

Measurement of ATP hydrolysis activity. ATP hydrolysis measurements were performed using a coupled enzyme assay as described (Laubinger and Dimroth 1988) with the following modifications. Instead of potassium phosphate, 50 mM Tris-Cl, pH 7.5 was used and Triton X-100 was omitted in the experiments with intact proteoliposomes.

Determination of buffer capacity of liposomes. The buffer capacity of the lipid headgroups was determined as described (Brune et al. 1987) with the following modifications. Liposomes containing Na^+ or K^+ were prepared as described above in 50 mM Na_2SO_4, 2.5 mM $MgCl_2$ and 50 mM K_2SO_4, 2.5 mM $MgCl_2$, respectively, diluted to 10 mg lipid/ml and adjusted to pH 6. Aliquots of a 10 mM KOH solution were added and the pH change was recorded with a glass electrode.

Proportion of empty and F_0 containing liposomes. The fraction of empty liposomes was determined as described (Franklin et al. 2004). Briefly, $\Delta\psi$-driven H^+-transport through F_0 was initiated by the addition of 8 nM valinomycin and followed until completion of the reaction. Then, 2 μl of a 2 mM stock solution of CCCP[1] was added to allow H^+-transport into empty

---
[1] Carbonyl cyanide *m*-chlorophenylhydrazone

2.3 Material and Methods

liposomes. The pH change observed before CCCP addition (F_0 liposomes) was divided by the total pH change (F_0 and empty liposomes) to obtain the fraction of liposomes containing an F_0 part.

Fluorescence measurements. Pyranine fluorescence was measured with a Cary Eclipse fluorescence spectrophotometer (Varian Inc., Palo Alto, USA). The excitation wavelengths were set to 405 and 460 nm and their emission at 510 nm was recorded with 2 Hz. The slits were set to 20 nm and the photomultiplier voltage was set to 550 V. During measurements the fluorescence spectrophotometer can remain open, which facilitates *in situ* addition of chemicals.

During kinetic measurements the following series of manipulations was performed if not otherwise stated. In a 5 ml plastic cuvette, 2.5 ml of buffer was mixed with 10 μl of the preincubated proteoliposome suspension. A baseline was recorded for 20 to 30 s, before the H^+-transport was initiated by the addition of 2 μl of a 10 μM valinomycin stock solution in ethanol and rapid mixing with a pipette. This mixing procedure proofed to be more effective than using a magnetic stirrer during measurements. All transport measurements were performed at room temperature.

Determination of the inner liposome volume. The inner volume of a liposome suspension was determined using pyranine as membrane impermeable fluorophore. Samples (1 ml) containing liposomes (30 mg lipid/ ml) in buffer C (buffer A supplemented with 50 mM KCl) were prepared in presence or absence of 1 mM pyranine (pyranine or control liposomes) as described above. The two liposome preparations were then diluted threefold with buffer C containing no or 1 mM pyranine to yield two suspensions containing the same total amount of pyranine. The liposomes were collected by centrifugation (200'000 g, 45 min, 4 °C) and resuspended in 1 ml of buffer C. The external pyranine was removed by gel filtration on a PD-10 column using buffer C. The turbid fractions were combined and the liposomes were collected by centrifugation (200'000 g, 45 min, 4 °C) and resuspended in 1 ml buffer C. The total amount of phospholipid was determined after centrifugation and thorough removal of excess of buffer by weighing the wet liposome fractions using a balance with an accuracy of 0.1 mg (Mettler Toledo, Greifensee, Switzerland). Typically, 30 mg of dry phospholipids displayed a total mass of 90-100 mg after hydration. The total amount of pyranine in the two preparations was determined by fluorescence spectroscopy in buffer C, supplied with 0.2 % Triton X-100 (to solubilize the liposomes). Samples of 4 μl of either liposome preparation were mixed with 2.5 ml buffer C and the fluorescence emission at 510 nm was measured (excitation wavelength was set to 412 nm). A standard curve using different amounts of pyranine (from a 10 μM stock solution) in buffer C with Triton X-100 was used to quantify the total amount of pyranine measured in the liposome samples.

2 Δψ and ΔpH are equivalent driving forces for the F_0 part

To determine the amount of incorporated pyranine, the fluorescence of the control liposomes was subtracted from the fluorescence signal of the pyranine liposomes. The inner volume was then calculated assuming an internal pyranine concentration of 1 mM.

Western Blot analysis. Western blots using antibodies against subunit β (from rabbit, AB-Cam, Cambridge, UK) and against subunit b (from mouse, described in Birkhenhager et al. (1995)) were used according to the protocol of the manufacturer. Secondary antibody incubation was performed with horseradish peroxidase conjugated to anti-rabbit and anti-mouse antibodies, respectively, and protein bands were visualized using the ECL detection system (GE Healthcare, Glattbrugg, Switzerland).

Mathematical fits. For the mathematical fits indicated, the program SigmaPlot (SyStat Software, Erkrath, Germany) was used.

2.4 Results and Discussion

2.4.1 Unidirectional Reconstitution

To analyze proton flux through the F_0 complex in a specific direction, proteoliposomes with a defined orientation of the F_0 complex are required. In a facile protocol for the reconstitution of *E. coli* ATP synthase, preformed liposomes are treated with Na^+-cholate and incubated with the purified enzyme. The detergent is removed subsequently by gel filtration (Ishmukhametov et al. 2005).

The procedure was optimized for our purposes as described in Material and Methods (see chapter 2.3) and the orientation of the reconstituted F_1F_0 ATP synthases was investigated by two independent methods. First, the enzyme orientation in the proteoliposomes was determined by measuring ATP hydrolysis activity. Since ATP is not membrane permeable, only the enzymes with the F_1 headpiece facing outwards contribute to ATP hydrolysis. After solubilization of the liposomes by a mild detergent, however, ATP is hydrolyzed by enzymes of either orientation. In a second procedure, the F_1 headpiece from outwards oriented enzymes was stripped off and removed by two subsequent washing steps with low ionic strength buffer containing EDTA. The efficiency of removing the external F_1 parts by this procedure was > 98 %. To determine the amount of inwards oriented enzymes, the liposomes were solubilized and ATP hydrolysis activity was determined. The results of both experiments (see table 2.1) show a unique orientation of the F_1F_0 ATP synthase with > 97 % of the F_1 moieties facing outwards. These conclusions were confirmed by Western blot analyses with antibodies against

2.4 Results and Discussion

Table 2.1: Orientation of ATP synthase after reconstitution into liposomes.
Liposomes containing *E. coli* F_1F_0 ATP synthase were prepared as described in Material and Methods. The specific ATPase activity was 18 U/ mg protein. Half of the liposomes were treated with low ionic strength buffer as described to strip off external F_1 headpieces to form F_0 liposomes. A sample of 20 µl of liposomes was then mixed with 500 µl ATP hydrolysis assay buffer and the activity was monitored at 340 nm by a coupled spectrophotometric assay as described (Laubinger and Dimroth 1988). CCCP (2 µM final concentration) was added to prevent the formation of an inhibitory potential and Triton X-100 (0.1 % final concentration) was added to solubilize the liposomes. The internal ATPase activity was calculated from the difference of activities before and after addition of Triton X-100.

Additions	Activity (mU/ mg lipid)	Rel. activity (%)	Rel. internal activity (%)
		F_1F_0 liposomes	
ATP	11.67	31.37	
CCCP	36.17	97.19	
Triton X-100	37.21	100	2.81
		F_0 liposomes	
ATP	0.40	1.08	
CCCP	0.61	1.64	
Triton X-100	1.32	3.56	1.92

subunit β and b of liposome preparations before and after stripping of the F_1 parts. The results of figure 2.1 show that the β-subunit was completely removed by the stripping procedure while the amount of subunit b remained unchanged (see figure 2.1, upper and lower panel). From these experiments, an entirely uniform orientation of F_0 parts was inferred.

2.4.2 Pyranine as indicator of internal pH change

Pyranine is a pH-dependent fluorophore with high polarity due to three sulfonate groups. The resulting membrane impermeability makes it particularly suitable for entrapment into lipid vesicles (for review, see Dencher et al. (1986)). Using an emission wavelength of 510 nm, the fluorophore was excited at 405 nm or 460 nm and the ratio of the respective emission intensities at 510 nm was calculated. This ratio displays a very reproducible dependency on the pH value in the range from pH 6 to pH 9 (Fig. 2.1 B). To measure pH changes inside liposomes, pyranine was entrapped into the F_0 proteoliposomes by two freeze/ thaw/ sonication cycles in the presence of 50 mM K_2SO_4 to enhance fluorophore uptake. The external pyranine was subsequently removed by gel filtration and the liposomes were concentrated by centrifugation. In initial H^+-transport experiments, the proteoliposomes were energized with a $\Delta\psi$ generated through a K^+/ valinomycin diffusion potential and the emission traces at 510 nm excited at 405 nm or 460 nm respectively, were recorded (Fig. 2.1 C). The course of the reaction is expressed as the ratio of

2 Δψ and ΔpH are equivalent driving forces for the F_0 part

the emission values between the two excitation wavelengths (Fig. 2.1 C, inset). The setup as described so far is sufficient for qualitative measurements in order to test the capability of a proteoliposome preparation to translocate protons. Within a single preparation of liposomes, the output is also quantitatively correct on a relative level (when no absolute transport rates are required). Effects of inhibitors or mutations can therefore be investigated at this level.

Measurement of the pH within proteoliposomes. For determination of absolute transport rates, the flux of protons between proteoliposomes and the environment had to be determined by measuring the pH within liposomes, while the external pH remained constant. The fluorescence emission of liposomes at either wavelength stems from pyranine in F_0 containing ($F_{in,p}$) and F_0-free liposomes ($F_{in,e}$) and from residual external pyranine (F_{ex}).

$$F_{tot} = F_{in,p} + F_{in,e} + F_{ex} \tag{2.1}$$

To determine the amount of external pyranine, the fluorescence of a liposome preparation (pH 7.2) in buffer with pH 7.2 was measured first, followed by a rapid shift of the external pH to ~ 6.8 after addition of 20 μl 50 mM sulfuric acid. This pH jump will only affect fluorescence of the external pyranine while that of the internal pyranine remains unchanged. This allowed determination of the fraction of internal pyranine (x_{in}) in a liposome preparation as described in Appendix I on page 107:

$$x_{in} = \frac{F_{tot(405,pH\,6.8)} - F_{tot(460,pH\,6.8)} \times R_{(pH\,6.8)}}{F_{tot(405,pH\,7.2)} - F_{tot(460,pH\,7.2)} \times R_{(pH\,6.8)}} \tag{2.2}$$

The terms $F_{tot(460,\,pH\,6.8)}$, $F_{tot(405,\,pH\,6.8)}$ and $F_{tot(460,\,pH\,7.2)}$, $F_{tot(405,\,pH\,7.2)}$ correspond to the measured fluorescence values of a liposome preparation (internal pH 7.2) in buffers with pH 6.8 and pH 7.2, respectively. The value for $R_{pH\,6.8}$ was obtained from figure 2.1 B. The absolute contribution of externally bound pyranine can be calculated with equations 2.3 and 2.4 for any given pH Y.

$$F_{ex\,(405,\,pH\,Y)} = F_{tot\,(405,\,pH\,Y)} - F_{tot\,(405,\,pH\,7.2)} \times x_{in} \tag{2.3}$$

$$F_{ex\,(460,\,pH\,Y)} = \frac{F_{ex\,(405,pH\,Y)}}{R_{pH\,Y}} = \frac{F_{tot\,(405,pH\,Y)} - F_{tot\,(405,pH\,7.2)} \times x_{in}}{R_{pH\,Y}} \tag{2.4}$$

The calculated amount of external fluorescence was subtracted from the measured fluorescence to obtain the total internal fluorescence at each wavelength. In addition, the internal fluorescence of F_0-free liposomes, whose internal pH does not change during the reaction, was subtracted. The amount n_e of F_0-free liposomes was determined as described in the methods section. Values of 50-60 % and 75-85 % F_0-free liposomes for preparations from *E. coli* and spinach chloroplast, respectively were obtained (Fig. 2.1 D). These considerations lead to the following time course for each wavelength.

$$F_{i(405)}(t) = F_{tot(405)}(t) - F_{ex(405)}(t_0) - F_{tot(405)}(t_0) \times n_e \qquad (2.5)$$

$$F_{i(460)}(t) = F_{tot(460)}(t) - F_{ex(460)}(t_0) - F_{tot(460)}(t_0) \times n_e \qquad (2.6)$$

The course of the reaction was expressed as function of time

$$r(t) = \frac{F_{i(405)}(t)}{F_{i(460)}(t)} \qquad (2.7)$$

Using the mathematical fit from figure 2.1 B, the ratio r(t) can be converted into the corresponding pH(t) function. For a solution containing 50 mM Na_2SO_4, the following term was obtained

$$pH(t) = 7.25 - 0.539 \times \ln\left(r(t) - 0.312\right); R^2 = 0.998 \qquad (2.8)$$

For a solution containing 50 mM K_2SO_4, we obtained

$$pH(t) = 7.14 - 0.559 \times \ln\left(r(t) - 0.295\right); R^2 = 0.997 \qquad (2.9)$$

Characterization of the experimental setup. A K^+/ valinomycin diffusion potential was applied as driving force for H^+-transport. In these experiments, the concentration of valinomycin must be high enough to ensure that K^+-transport by valinomycin is not rate limiting. Excessive valinomycin concentrations, however, which mediate H^+-transport must be avoided (Kríz et al. 2006). Therefore, a suitable valinomycin concentration has to be determined for each type of experiment. As depicted in figure 2.2 A, samples with final valinomycin concentrations of 80 and 800 pM clearly show a decreased uptake rate compared to samples with 8 and 16 nM valinomycin. On the other hand, only a minimal background H^+-transport was observed at 16 nM valinomycin when the ATP synthase was preincubated with 20 μM DCCD[2], a covalent inhibitor of the F_0 part. For further experiments a concentration of 8 nM valinomycin was used, if not otherwise indicated.

The integrity of the system was tested by incubating F_0 liposomes with isolated F_1 ATPase. After 45 min incubation at 4 °C, the liposomes were diluted into assay buffer and the sample energized with a $\Delta\psi$ and H^+-transport into the liposomes was determined (Fig. 2.2 B). H^+-transport was drastically reduced (< 10 %) after incubation of the liposomes with isolated

---

[2]Dicyclohexylcarbodiimide

2 Δψ and ΔpH are equivalent driving forces for the F_0 part

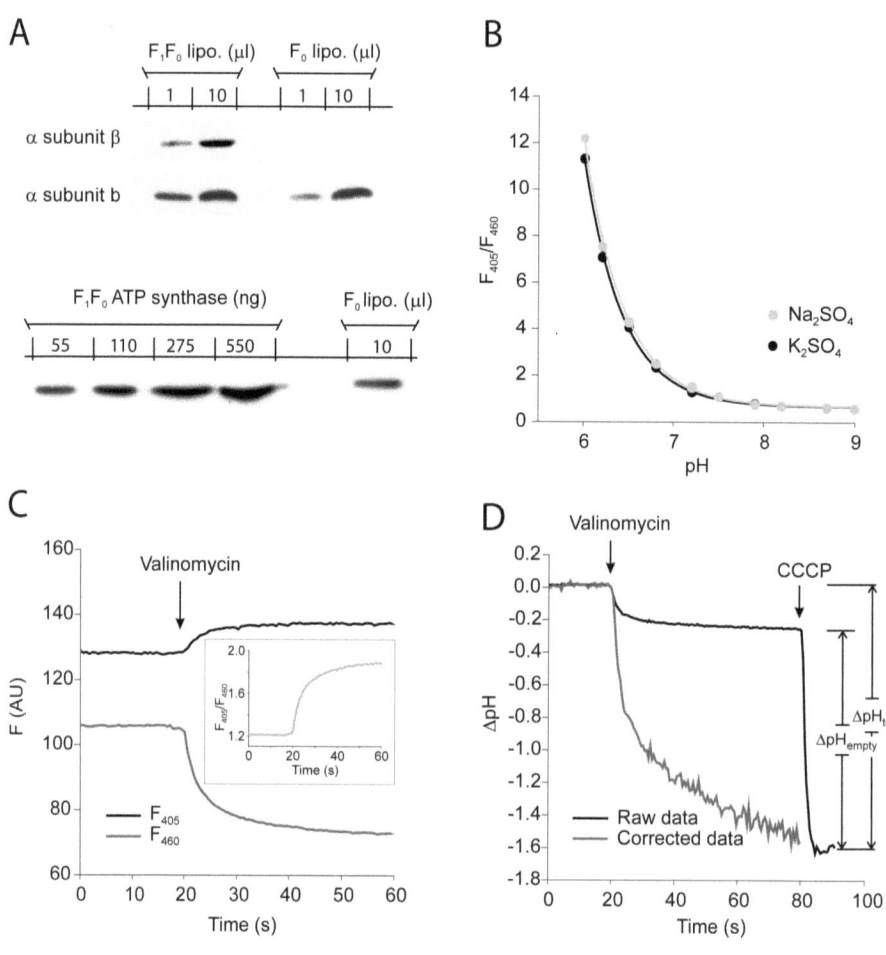

F_1, indicating that the F_1 parts had rebound to the membrane embedded F_0 parts to prevent H^+-transport.

2.4.3 Initial rates of H^+-transport through F_0 of *E. coli* or spinach chloroplasts

Maximal rates for ATP synthesis and hydrolysis have been described to be around 50 to 250 ATP × s⁻¹ (Etzold et al. 1997; Moriyama et al. 1991), corresponding to a maximal H^+-translocation rate of 150-1200 H^+/ (s × enzyme). For the isolated F_0 part, however proton transport rates have been found that varied drastically up to very high values (70 - 100000 H^+/ (s × enzyme)) (Cao et al. 2001; Lill et al. 1986; Sone et al. 1981), and only in recent years,

---

Figure 2.1 *(preceding page)*: **A. Western Blot analysis of liposome preparations (Upper panel).** *E. coli* F_1F_0 ATP synthase was reconstituted into liposomes and the F_1 part was stripped off as described (Ishmukhametov et al. 2005; Kluge and Dimroth 1992). Samples of 1 µl and 10 µl before and after the stripping procedure were subjected to SDS-PAGE, blotted onto nitrocellulose and subunits β and b were detected by Western blotting. **(Lower panel)** A serial dilution of purified F_1F_0 ATP synthase of *E. coli* was used to quantify the amount of F_0 within the proteoliposomes. Indicated are the amounts of purified ATP synthase as calculated from protein determination and the amount of liposomes used per measurement. Samples were subjected to SDS-PAGE, blotted onto nitrocellulose and subunit b was detected by Western blotting. Western blot were scanned and signal intensities were obtained using QuantityOne (BioRad, Reinach, Switzerland) **B. pH dependence of pyranine fluorescence.** The fluorescence emission at 510 nm, excited at either 405 nm or 460 nm, of 20 nM pyranine in buffer solution (2 mM buffer, 2.5 mM MglCl$_2$ and 50 mM Na$_2$SO$_4$/ K$_2$SO$_4$) was measured. The following chemicals were used to buffer at the indicated pH: MES (pH 6, 6.2, 6.5); MOPS (pH 6.8, 7.2); POPSO (pH 7.5, 7.9, 8.2, 8.4); CHES (pH 8.7, 9). The emission ratio F_{405}/F_{460} was then plotted against the pH values (circles) and a mathematical model using an exponential decay with three parameters was used to fit the correlation as indicated (line). **C. Original data obtained in the measurements.** In a 5 ml plastic cuvette, 5 µl to 10 µl of *E. coli* liposome solution (2 mM MOPS-NaOH, pH 7.2, 2.5 mM MgCl$_2$, 10 mM Na$_2$SO$_4$) was mixed with 2.5 ml of assay buffer (2 mM MOPS-NaOH, pH 7.2, 2.5 mM MgCl$_2$, 10 mM K$_2$SO$_4$) and florescence was recorded at 510 nm (excitation at 405 nm or 460 nm). After 20 to 30 s, H^+-transport was initiated by addition of 2 µl valinomycin from a 10 µM stock solution in ethanol and the sample was quickly mixed with a pipette. The ratio of the values (F_{405}/F_{460}) obtained were plotted against the time (inset). **D. Contribution of empty liposomes to the fluorescence signal.** Using proteoliposomes containing F_0 part of spinach chloroplasts, a similar time course as described in 2.1 C was recorded. After H^+-transport has reached a steady state level, 2 nM CCCP was added to the sample and quickly mixed with a pipette to induce H^+-transport of empty liposomes. The trace was corrected for the external pyranine and converted into ΔpH values as described in the Methods section. Assignments of total pH change (ΔpH_{tot}) and contribution of empty liposomes (ΔpH_{empty}) was used to calculate the amount of empty liposomes. The fluorescence values at F_{405} and F_{460} were then corrected by subtraction of the fluorescence contribution of empty liposomes and again, the ratio was calculated and converted into ΔpH (grey curve).

2 Δψ and ΔpH are equivalent driving forces for the F_0 part

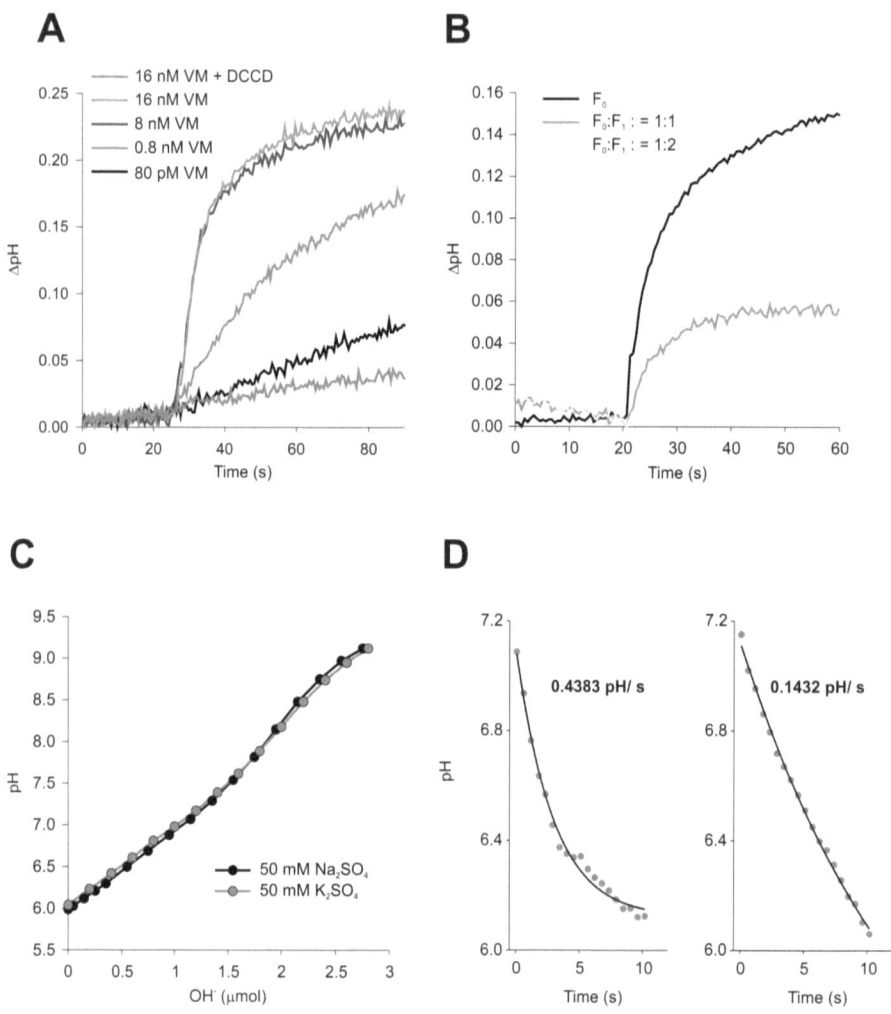

2.4 Results and Discussion

comparable rates of 3000 to 8000 H^+/s were obtained (Feniouk et al. 2004; Franklin et al. 2004). The experimental systems were quite complex and diverse and therefore, the data are difficult to compare. When the transport rate was measured in small particles of *Rhodobacter capsulatus*, the system was energized with flash induced H^+-transport by photosynthetic reaction centers and the relaxation of the induced potential was measured following the absorption changes of intrinsic chromophores (Feniouk et al. 2004). The measurements were highly time-resolved but could not unequivocally establish that all the observed processes can be attributed to H^+-transport through F_0 parts and not to other transport proteins present in the system. Furthermore, rapid H^+-flux by primary proton pumps across the membrane has been proposed to generate a local proton gradient that exists transiently before the translocated protons equilibrate with the bulk (Heberle et al. 1994). Since only fast relaxations (<100 ms) of the induced potential were measured, transient proton gradients cannot be neglected, as they might present an additional driving force for H^+-transport through F_0 parts. Hence, there is some uncertainty in the actual size of the applied driving force, which has been discussed in part.

These problems can be overcome with artificial energization by a K^+/valinomycin diffusion gradient. This has been applied to determine the proton transport rate through F_0 of the *E. coli* ATP synthase (Cao et al. 2001; Franklin et al. 2004; Sone et al. 1981). In these studies the pH change during H^+-transport was measured with a sensitive pH electrode in the external buffer medium (Greie et al. 2000; Sone et al. 1981). Since the effective pH change is very small, a disadvantage of this method is a rather low signal-to-noise ratio. In our experiments the

---

Figure 2.2 *(preceding page)*: **A. Effect of valinomycin concentration on initial transport rates.** To test the effect of the valinomycin concentration, $\Delta\psi$-driven H^+-transport in synthesis direction was measured with the enzyme of *E. coli* (efflux from liposomes). Experimental conditions were similar as described in figure 2.1 C, with the exception, that different amounts of valinomycin were added to initiate proton transport as indicated. As a negative control, DCCD treated liposomes were mixed with the highest concentration of valinomycin (red trace). **B. Inhibition of F_0 H^+-transport after rebinding of soluble F_1.** *E. coli* F_0 liposomes (2 mM MOPS-NaOH, pH 7.2, 2.5 mM $MgCl_2$, 50 mM Na_2SO_4, 0.125 mM K_2SO_4) were mixed with purified soluble F_1 part in the molar ratios indicated and incubated at 4 °C for 45 min. The liposomes were then diluted into K^+ assay buffer (2 mM MOPS-NaOH, pH 7.2, 2.5 mM $MgCl_2$, 50 mM Na_2SO_4, 0.125 mM K_2SO_4) and $\Delta\psi$-driven efflux was initiated by the addition of 8 nM valinomycin. **C. Determination of buffer capacity of the liposome preparation.** Liposomes prepared in either 50 mM Na_2SO_4, 2.5 mM $MgCl_2$ or 50 mM K_2SO_4, 2.5 mM $MgCl_2$ were diluted to 10 mg lipid/ml and adjusted to pH 6.0. Aliquots of a 10 mM KOH solution were added and the pH change was recorded with a glass electrode. **D. Determination of the initial rate of H^+-transport.** Shown are experiments of $\Delta\psi$-driven H^+-transport in hydrolysis direction for the *E. coli* (left) and the spinach chloroplast (right) enzymes. The ten first seconds after valinomycin addition were used to fit the experimental data (red circles) mathematically (black curve, fitted as exponential decay with 3 parameters) and the initial rate was calculated from the first derivative.

2 Δψ and ΔpH are equivalent driving forces for the F_0 part

disadvantages of contaminating enzymes or low signal-to-noise ratio were avoided by reconstituting the system from purified components and following the pH changes within the liposomes by a sensitive optical method. In the following sections, the requirements for the determination of H^+-transport rates of the ATP synthases of *E. coli* and spinach chloroplasts are described.

Buffer capacity of the liposomes. The buffer capacity of the phospholipid headgroups lining the lumen of the liposomes exceeds by far the buffering capacity of the internal buffer (2 mM MOPS-NaOH) (Brune et al. 1987; Dencher et al. 1986), but varies with the lipid composition. Therefore the buffering capacity of the employed liposome preparation was determined by titration with KOH as described in Material and Methods (Fig. 2.2 C). The buffering capacity was found to be approximately linear over pH segments of ~ 0.5 pH units and displays a value of ~ 100 µmol H^+/ (g soy bean lipid × pH unit) in the neutral range, which is in good agreement with similar experiments (Brune et al. 1987; Dencher et al. 1986). The absolute buffer capacity can be calculated from the amount of phospholipid used per measurement. The contribution of the internal buffer to the buffer capacity is small (~ 1 - 5 %), but can be calculated from the Henderson-Hasselbalch equation together with the internal volume of the liposomes. The internal volume was found to be 1.5 to 1.8 µl/ mg lipid by pyranine fluorescence as described in Material and Methods. Additionally, the liposome preparation was analyzed using electron microscopy (data not shown). The size distribution of the liposomes roughly ranged from 40 to 120 nm diameter which corresponds to an internal volume of 0.27 - 7.2 × 10^{-18} l per liposome. If an average mass of 700 Da and an average surface area of 70 $Å^2$ is assumed for a single lipid molecule, a volume of 2 - 6 µl/ mg lipid results. This estimation is slightly higher than the value obtained with the pyranine measurement but within the same range.

2.4.4 Number of functional F_0 molecules

A major difficulty in determining the transport rate per single F_0 complex is the determination of the number of active F_0 complexes in a measurement. Assuming that every liposome contains either one or no F_0 complex, the number of active F_0 complexes can be calculated from the total number of liposomes and the fraction of liposomes contributing to H^+-transport (Feniouk et al. 2004; Franklin et al. 2004). A weakness of this approach is the exact determination of the total number of liposomes which is very difficult. It can either be calculated from the inner volume of the liposomes (Brune et al. 1987) or the average liposome size (from electron microscopy or dynamic light scattering data) (Franklin et al. 2004). Thereby, the number is proportional to the volume of the liposomes. Since the volume is proportional to r^3, a little change in size determination drastically alters the number of liposomes. Alternatively, the amount of F_0-

subunits in our preparation was quantified by Western Blot analysis against subunit b. This allowed the calculation of the maximal amount of incorporated F_0 parts and the subsequent rate determination reflects a minimal value. Using a serial dilution of highly purified protein, an amount of 3.87 fmol of *E. coli* and 1.72 fmol of spinach chloroplast ATP synthase was found to have been incorporated per mg of lipid, respectively (Fig. 2.1 A, lower panel). This reflected an overall reconstitution efficiency of $\sim$ 13 %.

Determination of initial rates. The experimental time courses were fitted mathematically to obtain the initial slopes as shown in figure 2.2 D. From these data, the total number of translocated protons was calculated, taking into account the buffer capacity of the liposomes and the internal buffer. If the contribution of the internal buffer ($\sim$ 1 - 5 %) was neglected, rates could alternatively be calculated without the knowledge of the internal volume. The maximal number of F_0 parts present in the reaction sample was calculated based on the amount of lipid per measurement. A typical calculation is exemplified in Appendix II on page 109. With these considerations, for the enzyme from *E. coli*, energized solely with a membrane potential of 120 mV, a minimal rate of H^+-transport of 2400 H^+/ (s $\times$ F_0 part) was obtained. Similarly for the enzyme of spinach chloroplast, a minimal rate of 6300 H^+/ (s $\times$ F_0 part) was obtained.

2.4.5 $\Delta\psi$- and ΔpH-driven H^+-transport in synthesis and hydrolysis direction

Proton uptake into unidirectional F_0 liposomes prepared by the method described above reflects the flux of protons in ATP hydrolysis direction. To measure the reverse flux of protons reflecting the ATP synthesis direction, a reverse $\Delta\psi$ (positive inside) was applied and the increase of the internal pH due to proton efflux was followed by the pyranine fluorescence technique. In addition, the effect of the two parameters of the proton-motive force ΔpH and $\Delta\psi$ on the H^+-transport rate through F_0 was compared. To prevent the formation of a counteracting membrane potential in experiments with ΔpH-driven H^+-transport, 50 mM K_2SO_4 and valinomycin were present on both sides of the liposomes. Conversely, liposomes energized by $\Delta\psi$ were incubated over night at the corresponding pH to assure that no ΔpH was present. To compare the effect of each driving force on the transport rates, a constant driving force of 73 mV, consisting either of a ΔpH = 1.2 or a $\Delta\psi$ (K^+_{in}/K^+_{out} ratio of 15.8) was applied. As control served an inactive mutant of the *E. coli* ATP synthase, where the essential cD61 was replaced by an asparagine (Hoppe et al. 1982). When these liposomes were energized with $\Delta\psi$ no H^+-translocation could be determined, reinforcing the tightness of the proteoliposome preparation. Curiously when ΔpH was the sole driving force in these control liposomes, a slight drift of the

2 $\Delta\psi$ and ΔpH are equivalent driving forces for the F_0 part

signal was observed in the absence of valinomycin, indicating a slow H^+-translocation. The drift was similarly observed, when the F_0 part was labeled with the covalent inhibitor DCCD and therefore most likely reflects unspecific H^+-transport across the membrane. The results of selected H^+-transport measurements are depicted in figure 2.3 A-D. In synthesis direction the

Table 2.2: Initial H^+-transport through F_0.
Initial transport rates are summarized from figures 2.2 D and 2.3 A-D and from the literature.

Organism	Driving force	Direction	$H^+/$ (s × F_0 part)	Reference
E. coli	$\Delta\psi = 108$ mV	n. d.	3100	Franklin et al. (2004)
Bacillus PS3	$\Delta\psi = 94$ mV	Hydrolysis	47 (at 25 °C)	Sone et al. (1981)
R. capsulatus	$\Delta\psi = 100$ mV	Synthesis	6240	Feniouk et al. (2004)
Chloroplast	variable	Synthesis	Up to 10000	Lill et al. (1986)
E. coli	$\Delta\psi = 120$ mV	Hydrolysis	2351	This study
Chloroplast	$\Delta\psi = 120$ mV	Hydrolysis	6300	This study
E. coli	$\Delta\psi = 73$ mV	Synthesis	641	This study
	$\Delta\psi = 73$ mV	Hydrolysis	1358	This study
	$\Delta pH = 73$ mV	Synthesis	563	This study
	$\Delta pH = 73$ mV	Hydrolysis	1399	This study
Chloroplast	$\Delta\psi = 73$ mV	Synthesis	2389	This study
	$\Delta\psi = 73$ mV	Hydrolysis	3683	This study
	$\Delta pH = 73$ mV	Synthesis	3117	This study
	$\Delta pH = 73$ mV	Hydrolysis	3530	This study

total amount (not the rate) of transported protons was bigger in the presence of ΔpH compared to $\Delta\psi$ while both driving forces were comparable in hydrolysis direction. This effect is likely to be related to the mode of $\Delta\psi$ generation. Transport in synthesis direction requires an inside positive $\Delta\psi$, which is obtained by a low internal and high external K^+ concentration. After valinomycin addition, K^+ ions flow into the liposomes and lead to a rapid rise of K^+ concentration due to the small internal volume of the liposomes. This again diminishes the $\Delta\psi$ and thus the driving force. This effect is less pronounced, when an inverse $\Delta\psi$ is generated and a high K^+ concentration is present within the liposomes, since the external concentration can be assumed to remain constant. For comparison of the different driving forces in each specific direction, the initial transport rates were calculated as described above and summarized in table 2.2. For the enzyme from *E. coli*, H^+-transport in hydrolysis direction was found to be twice more efficient compared to the synthesis direction. Within one direction, the driving forces produced similar rates. The difference between hydrolysis and synthesis was less obvious in the spinach chloroplast enzyme, although slightly higher rates were obtained in the hydrolysis direction. Interestingly, ΔpH, which is the predominant driving force in chloroplasts, was found to be more efficient in driving proton transport in synthesis direction. Generally, rates at similar driving forces were up to three times higher in the chloroplast F_0 compared to the E. coli F_0 part. This is in good corroboration with the maximal rates of 600 ATP × s^{-1} and 240 ATP × s^{-1} in the respective holoenzymes (Engelbrecht et al. 1989; Etzold et al. 1997).

These findings further indicate a certain difference of the chloroplast and the E. coli enzyme in their efficiency to utilize the different driving forces. It is likely, that these differences are even more prominent in the F_1F_0 holoenzyme.

2.4.6 Ohmic conductance of the E. coli F_0 part

In recent years, the F_0 parts of the ATP synthases of Bacillus PS3 (hydrolysis direction) and R. capsulatus (synthesis direction) were shown to exhibit a linear dependence of the H$^+$-transport rate on the driving force applied (Feniouk et al. 2004; Sone et al. 1981). In contrary to the Na$^+$-translocating F_0 part of the ATP synthase of Propionigenium modestum, no voltage threshold was observed in the H$^+$-translocating enzymes (Kluge and Dimroth 1992). We therefore utilized the described assay to investigate the kinetic properties of the E. coli F_0 part at pH 7.2. Na$^+$-liposomes (containing 50 mM Na$_2$SO$_4$, 0.5 mM K$_2$SO$_4$) and K$^+$-liposomes (containing 50 mM K$_2$SO$_4$) were prepared and used for a series of $\Delta\psi$-driven efflux and influx measurements, respectively. The size of the membrane potential was modulated by varying the external K$^+$-concentration according to the Nernst equation. As depicted in figure 2.4 A, linear relationships were obtained for both transport direction and no significant voltage thresholds were observed. As expected from the data in figure 2.3, transport in hydrolysis direction was roughly twice as efficient as in synthesis direction. A similar picture was obtained when ΔpH was the driving force for proton transport (Fig. 2.4 B). In these experiments, K$^+$-liposomes (internal pH 7.2) were measured in K$^+$ containing buffers with pH values varying from pH 6 to 9, and H$^+$-transport was initiated with the addition of valinomycin. These data show that and pH are equivalent driving forces in the F_0 part of the E. coli ATP synthase. It remains elusive, whether the different findings in the F_1F_0 holoenzyme have to be attributed to the F_1 part or to altered F_0 properties after tight coupling with the F_1 part.

2.4.7 Inhibition of H$^+$-translocation by tributyltin chloride

Tributyltin chloride (TBT) is a known inhibitor of oxidative phosphorylation in general and ATP synthases in particular. While it has been shown to specifically bind to subunit a of the ATP synthase of Ilyobacter tartaricus (von Ballmoos et al. 2004), no studies are reported about the interaction of TBT with purified F_0 during H$^+$-translocation. Therefore the ability of TBT to inhibit H$^+$-transport through the F_0 part of E. coli and spinach chloroplasts was investigated.

2 Δψ and ΔpH are equivalent driving forces for the F_0 part

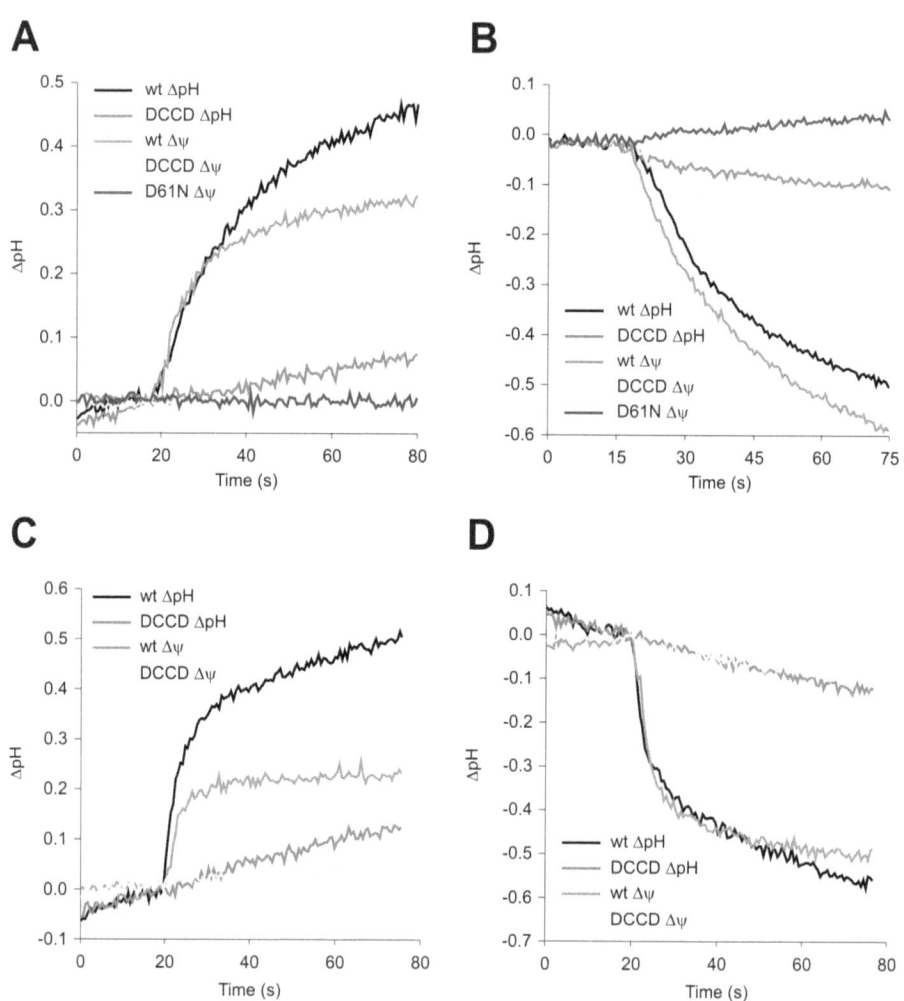

2.4 Results and Discussion

When TBT was added from a stock solution to a final concentration of 500 nM to the proteoliposomes in the assay buffer and incubated for 45 s, before the reaction was started with valinomycin, almost no inhibition was observed. The experiment was repeated with a DCCD inhibited enzyme, and surprisingly, TBT stimulated a pH change indicative of an unknown transport, which is not related to the F_0 part. At elevated concentrations, TBT is also known to act as an anion/ hydroxide exchanger. Since only very little lipid was present in our measurement, accumulation of TBT in the liposome is likely to exceed this critical concentration and transport of hydroxide anions might generate the observed pH change. On the other hand it is possible, that a certain TBT concentration is necessary to inhibit the ATP synthase (an apparent K_i of 200 nM has been determined for the *E. coli* enzyme in native membranes (von Ballmoos et al. 2004). Therefore a large excess of empty liposomes containing no pyranine was added to the sample to prevent accumulation of TBT in the F_0 liposomes. Indeed, a decreased uptake rate was observed compared to the first experiment. However, in the DCCD treated liposomes, TBT still stimulated a change of the internal pH. In the next series of experiments with DCCD treated liposomes, the valinomycin concentration was decreased stepwise until a significant pH change was no longer observed (data not shown). It appears that a combination of valinomycin and tribtutyltin chloride somehow enables unspecific ion transport across lipid bilayers. The experiments with and without TBT were then repeated in the presence

---

Figure 2.3 *(preceding page)*: A. H^+-transport in synthesis direction in E. coli liposomes. In all measurements, a constant driving force of 73 mV in the form of either a $\Delta\psi$ or a ΔpH was applied. For $\Delta\psi$-driven H^+-transport, 5 µl Na^+-liposomes (2 mM MOPS-NaOH, pH 7.2, 2.5 mM $MgCl_2$, 50 mM Na_2SO_4, 1 mM K_2SO_4) were mixed with 2.5 ml assay buffer (2 mM MOPS-NaOH, pH 7.2, 2.5 mM $MgCl_2$, 34 mM Na_2SO_4, 16 mM K_2SO_4) and H^+-transport was initiated with 8 nM valinomycin. For ΔpH-driven H^+-transport, 5 µl liposomes (2 mM MOPS-NaOH, pH 7.2, 2.5 mM $MgCl_2$, 50 mM K_2SO_4) were mixed with 2.5 ml assay buffer (2 mM POPSO-NaOH, pH 8.4, 2.5 mM $MgCl_2$, 50 mM K_2SO_4) and H^+-transport was initiated with 8 nM valinomycin. Initial rates resulting from these transport measurements were calculated as described in figure 2.2 D and are summarized in table 2.2. **B. H^+-transport in hydrolysis direction in *E. coli* liposomes.** In all measurements, a constant driving force of 73 mV in form of either a $\Delta\psi$ or a ΔpH was applied. For $\Delta\psi$-driven H^+-transport, 5 µl K^+-liposomes (2 mM MOPS-NaOH, pH 7.2, 2.5 mM $MgCl_2$, 50 mM K_2SO_4) were mixed with 2.5 ml assay buffer (2 mM MOPS-NaOH, pH 7.2, 2.5 mM $MgCl_2$, 46.875 mM Na_2SO_4, 3.125 mM K_2SO_4) and H^+-transport was initiated with 8 nM valinomycin. For ΔpH-driven H^+-transport, 5 µl liposomes (2 mM MOPS-NaOH, pH 7.2, 2.5 mM $MgCl_2$, 50 mM K_2SO_4) were mixed with 2.5 ml assay buffer (2 mM MES-NaOH, pH 6.0, 2.5 mM $MgCl_2$, 50 mM K_2SO_4) and H^+-transport was initiated with 8 nM valinomycin. Initial rates resulting from these transport measurements were calculated as described in figure 2.2 D and are summarized in table 2.2. **C. H^+-transport in synthesis direction in spinach chloroplast liposomes.** As 2.3 A, but liposomes containing spinach chloroplast ATP synthase were used. **D. H^+-transport in hydrolysis direction in spinach chloroplast liposomes.** As 2.3 B, but liposomes containing spinach chloroplast ATP synthase were used.

2 $\Delta\psi$ and ΔpH are equivalent driving forces for the F_0 part

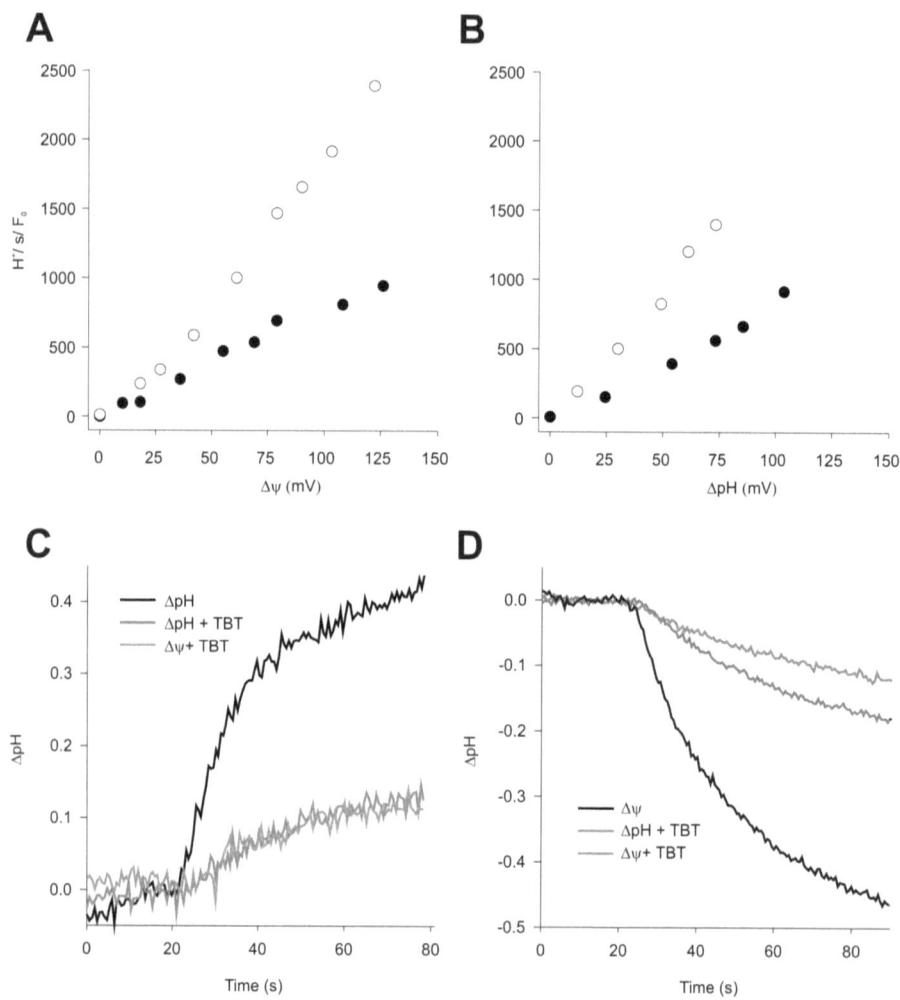

2.4 Results and Discussion

of additional lipids and a decreased valinomycin concentration of 1.2 nM. Sets of inhibition experiments with the enzymes of spinach chloroplast and *E. coli* are shown in figures 2.4 C and D, respectively. A significant decrease of H^+-transport was obtained in the presence of 500 nM TBT, irrespective of transport direction or driving force. It has to be considered that transport rates in the untreated enzyme are decreased due to the rate limiting effect of valinomycin as shown in figure 2.1 C and therefore the effective extent of TBT inhibition is even more pronounced.

Figure 2.4 *(preceding page)*: A. Effect of membrane potential on H^+-transport Na^+-liposomes containing *E. coli* F_0 part (2 mM MOPS-NaOH, pH 7.2, 2.5 mM $MgCl_2$, 50 mM Na_2SO_4, 0.5 mM K_2SO_4) were used to measure $\Delta\psi$-driven H^+-transport in synthesis direction (closed circles). The size of the membrane potential was calculated according to the Nernst equation and the K^+ concentration in the assay buffer was varied by mixing K^+ assay buffer (2 mM MOPS-NaOH, pH 7.2, 2.5 mM $MgCl_2$, 50 mM K_2SO_4) and Na^+ assay buffer (2 mM MOPS-NaOH, pH 7.2, 2.5 mM $MgCl_2$, 50 mM Na_2SO_4, 0.5 mM K_2SO_4) appropriately. Accordingly, Na^+-liposomes (2 mM MOPS-NaOH, pH 7.2, 2.5 mM $MgCl_2$, 50 mM K_2SO_4) were used to measure $\Delta\psi$-driven H^+-transport in hydrolysis direction (open circles). Plotted are the initial rates as calculated in figure 2.2 D. **B. Effect of proton gradient on H^+-transport** K^+-liposomes containing *E. coli* F_0 part (2 mM MOPS-NaOH, pH 7.2, 2.5 mM $MgCl_2$, 50 mM K_2SO_4) were used to measure ΔpH-driven H^+-transport in synthesis (closed circles) and hydrolysis direction (open circles). The liposomes were mixed with assay buffer at different pH values and H^+-transport was initiated by addition of 8 nM valinomycin. The following chemicals were used to buffer at the desired pH values: MES (pH 6, 6.2, 6.5); MOPS (pH 6.8, 7.2); POPSO (pH 7.5, 7.9, 8.2, 8.4); CHES (pH 8.7, 9). Plotted are the initial rates as calculated in figure 2.2 D. **C. Inhibition of H^+-transport in synthesis direction by TBT** Shown are the results of the F_0 part of the spinach chloroplast ATP synthase. The F_0 part of *E. coli* produced similar results. An aliquot of 20 µl of empty liposomes (30 mg/ml in the respective assay buffer, containing no enzyme and no pyranine) was mixed with 2.5 ml of assay buffer and TBT was added to a final concentration of 500 nM. K^+-liposomes were then added as described and the baseline was recorded after 45 s incubation. H^+-transport was initiated with 1.2 nM valinomycin. **D. Inhibition of H^+-transport in hydrolysis direction by TBT** Shown are the results of the F_0 part of the *E. coli* ATP synthase. The F_0 part of spinach chloroplast produced similar results. The experiment was carried out as described in figure 2.4 D, except that Na^+-liposomes were used for $\Delta\psi$-driven H^+-transport in hydrolysis direction.

2 $\Delta\psi$ and ΔpH are equivalent driving forces for the F_0 part

2.5 Concluding remarks

Here we describe a reliable system for determination of H^+-transport through F_0 parts from various organisms. The formation of F_0 liposomes from purified components eliminates the possibility of unspecified ion fluxes through contaminating protein components. The reconstitution procedure yields a homogeneous orientation of the enzyme in the liposome bilayers with the cytoplasmic part of the enzyme facing outwards. The proteoliposomes were very tight towards unspecific proton diffusion or compensating charge transport. To monitor pH changes during H^+-transport, pyranine was entrapped within the proteoliposomes. Pyranine is very hydrophilic and no leakage from entrapped molecules was observed. When measured ratiometrically, the fluorescence changes showed a very good correlation with the H^+ concentration in the range from pH 6 to pH 9. These properties make it a largely superior probe compared to ACMA, which is hardly quantitative and only allows measurement of an intravesicular pH decrease, but not pH increase. An additional advantage of pyranine is its relative ease of quantification, if the external pyranine is calculated and subtracted. Unlike to the measurement of pH changes in the external medium, internal pyranine measurements display a very high signal-to-noise ratio, which allows economic use of proteoliposomes for measurements. The described system allowed for the first time to monitor and quantify H^+-transport through the ATPase F_0 part, under clearly defined conditions in both transport directions with each driving force. Controls with DCCD inhibited or inactive enzyme displayed activities < 5 % and < 3 % respectively. In order to establish or to compensate for an electric membrane potential, valinomycin was present in all measurements. No significant unspecific H^+-transport by valinomycin as reported by others was observed (Franklin et al. 2004) under the experimental conditions applied. Certain care has to be taken, when TBT is used as an inhibitor, since it exhibits anion/ hydroxide exchanger properties, but the problem can be circumvented by addition of empty liposomes and usage of decreased valinomycin concentrations. The advantages and properties of the described experimental setup makes it particularly suitable for an extensive investigation of the utilization of different driving forces by the F_0 part of ATP synthases from different origins, that is currently in progress in our laboratory.

3 Glimpse into an ATP synthase's F_0 motor at work

3.1 Abstract

The electrochemical potential of H^+ or Na^+ created during the degradation of nutrients is used by the F_1F_0 ATP synthase to synthesize ATP, the universal energy currency of living cells. Intersubunit rotation elicited during ion translocation through the membrane-embedded F_0 part causes conformational changes in the water exposed F_1 part enabling ATP synthesis. In reverse, the enzyme acts as an ATP-driven ion pump creating the membrane potential which is critical for cell viability. While the F_1 part is understood in great detail, no equivalent knowledge exists for the F_0 part. Here we show, that the F_0 parts of a Na^+ and an H^+-dependent enzyme display major asymmetries with respect to their mode of operation. This is reflected by the requirement of $\sim$ 100 times higher Na^+ or H^+ concentrations for the synthesis than the hydrolysis of ATP. A similar asymmetry is observed during ion transport through isolated F_0 parts, indicating different affinities for the binding sites in the a/ c interface. Together with further data, we propose a model, in which the binding affinity of a site is modulated by the stator arginine, switching from low to high as it moves from the periplasmic to the cytoplasmic ion path. Our findings have important consequences for the mechanism of ATP synthesis *in vivo*. We identify a discrepancy between *E. coli* growth by oxidative phosphorylation at pH 7 - 8 and the necessity of pH < 6.5 for ATP synthesis *in vitro*. This lends support to the long standing hypothesis that during respiration, lateral proton diffusion leads to significant acidification at the membrane surface. These observations may stimulate an in depth investigation of lateral proton diffusion and its significance for ATP synthesis or other bioenergetic membrane reactions *in vivo*.

3.2 Glimpse into an ATP synthase's F_0 motor at work

Implicit in Mitchell's chemiosmotic model is the assumption that membrane potential and transmembrane ion gradients are equivalent driving forces for ATP synthesis (Mitchell 1979). This view is reinforced by experiments showing that the chloroplast ATP synthase is driven almost completely by ΔpH (Fischer and Gräber 1999), while the mitochondrial and bacterial ATP synthases operate on a combination of membrane potential and ΔpH (Sone et al. 1977). ATP synthesis experiments with the Na$^+$ ATP synthase of *P. modestum* and the H$^+$ ATP synthase of *E. coli*, however, established an essential role for the membrane potential (Fischer and Gräber 1999; Kaim and Dimroth 1999). This sheds doubt on the perfect equivalence of the two driving forces. Since the specific demand for either of these forces has key consequences for the enzyme mechanism we extended earlier studies with the ATP synthases of *P. modestum* and *E. coli*. In the course of these investigations unprecedented critical parameters for ATP synthesis and ion translocation were discovered.

When lipid vesicles containing the purified *P. modestum* ATP synthase were energized by a K$^+$/ valinomycin diffusion potential of $\sim$ 160 mV and varying ΔpNa values, ATP synthesis increased steadily with increasing ΔpNa above a threshold of $\sim$ 30 mV (Figure 3.1 A. When ΔpNa was kept constant around 100 mV and the $\Delta\psi$ was varied from 0 to 180 mV, efficient ATP synthesis required a $\Delta\psi$ > 100 mV, and a total driving force of > 170 mV (Figure 3.1 B). These data extend earlier findings, establishing ΔpNa as indispensable driving force (Kaim and Dimroth 1999). Further experiments were performed by maintaining ΔpNa and $\Delta\psi$ at 90 and 140 mV, respectively, and varying the Na$^+$ concentrations on both sides of the liposome membrane (Figure 3.1 C). The results show that for efficient ATP synthesis, elevated Na$^+$ concentrations (K$_D$ $\sim$ 35 mM) are required at the source *P*-side, independent of the size of the applied $\Delta\psi$ (Table 3.1). The maximal synthesis rate was 15 ATP/ (s × enzyme), which is tenfold higher than in previous reports, where lower Na$^+$ concentrations were used (Kaim and Dimroth 1999). ATP hydrolysis is also Na$^+$-dependent, but requires about 70-fold lower concentrations (K$_D$ $\sim$ 0.5 mM) than ATP synthesis. In accordance with these data, asymmetric Na$^+$ binding affinities were observed for the opposing transport directions through the isolated

Table 3.1: **Apparent K$_D$ values for Na$^+$ during ATP synthesis at different electrical potentials.**
Experimental procedures are similar as described in Figure 3.1 C. The ΔpNa was kept constant at 108 mV during all measurements. The data were fitted using a model for single ligand binding.

$\Delta\psi$	100 mV	120 mV	140 mV	160 mV
app. K$_D$	39 ± 5 mM	34 ± 6 mM	35 ± 3 mM	39 ± 5 mM

3.2 Glimpse into an ATP synthase's F_0 motor at work

F_0 part (Fig. 3.5 on page 59).

In contrast to its requirement for ATP synthesis, ΔpNa was not effective to drive Na^+-transport through F_0 (Kluge and Dimroth 1992). In these experiments, however, transport in ATP hydrolysis direction has been measured. When we modified the system and measured Na^+-transport in ATP synthesis direction, ΔpNa or $\Delta\psi$ were efficient driving forces with an additive effect if applied together, showing that ATP synthesis and ion transport in the proper direction respond similarly to the applied driving forces (Figure 3.1 D). Stimulated by these results we extended our investigations to the *E. coli* ATP synthase. The purified enzyme was reconstituted into phospholipid vesicles, which were energized by ΔpH through acid-base treatment and by $\Delta\psi$ with a K^+/ valinomycin diffusion potential. In ATP synthesis experiments with pH 5.75 at the *P*-side and varying pH values at the *N*-side, a ΔpH of 1 had to be present at minimum (Fig. 3.2 A+B). At a constant ΔpH (and thus a constant driving force), the synthesis rate decreased continuously by increasing the *P*-side pH from 5.75 to 7.0, where synthesis was completely abolished (Fig. 3.2 C). This finding is remarkable, since ATP hydrolysis is not significantly affected by the pH between 6.5 and 9 (von Ballmoos and Dimroth 2007). To investigate whether the observed asymmetry is apparent in the isolated F_0 part, H^+-transport was measured in each specific direction. (Fig. 3.2 D). In hydrolysis direction, the F_0 liposomes catalyzed $\Delta\psi$-driven proton transport between pH 6 and 9 at a constant rate, resembling the ATP hydrolysis pH profile. H^+-transport in ATP synthesis direction, however, decreased continuously with increasing pH values from 6 to 9 to less than 20 % of the initial rate. Thus, the pH profiles for ATP synthesis and H^+-conduction in the appropriate direction resemble each other except for a shift of the apparent pK_a from 6.2 to 7.7. This shift may be related to the demand for increased torque during ATP synthesis, where counteracting conformational restraints in F_1 have to be overcome. The F_0 part of the *P. modestum* and the *E. coli* ATP synthase show major asymmetric properties during operation in ATP synthesis or hydrolysis direction. Particularly interesting is the requirement of 50-500 fold higher coupling ion concentrations at the source side for operation in ATP synthesis direction. These data may indicate different affinities of the binding site at the periplasmic and cytoplasmic entrance route of the ion, as postulated in theoretical F_0 motor models (Elston et al. 1998; Feniouk et al. 2004) or might result from an additional low affinity site in the periplasmic access route. In the first case, a mutation at the binding site is expected to equally affect both affinities, whereas in the second it should mainly affect the high binding affinity. We probed these possibilities with the cY70F mutant of the *I. tartaricus* enzyme and found highly elevated Na^+ concentration requirements for both ATP hydrolysis and synthesis ($K_D \sim$ 33 mM and > 190 mM, respectively) (Fig. 3.3), in support for the first possibility. In summary, we conclude that the affinity of the binding site decreases by about 2 orders of magnitude as it moves from the cytoplasmic to the periplasmic entrance port

3 Glimpse into an ATP synthase's F_0 motor at work

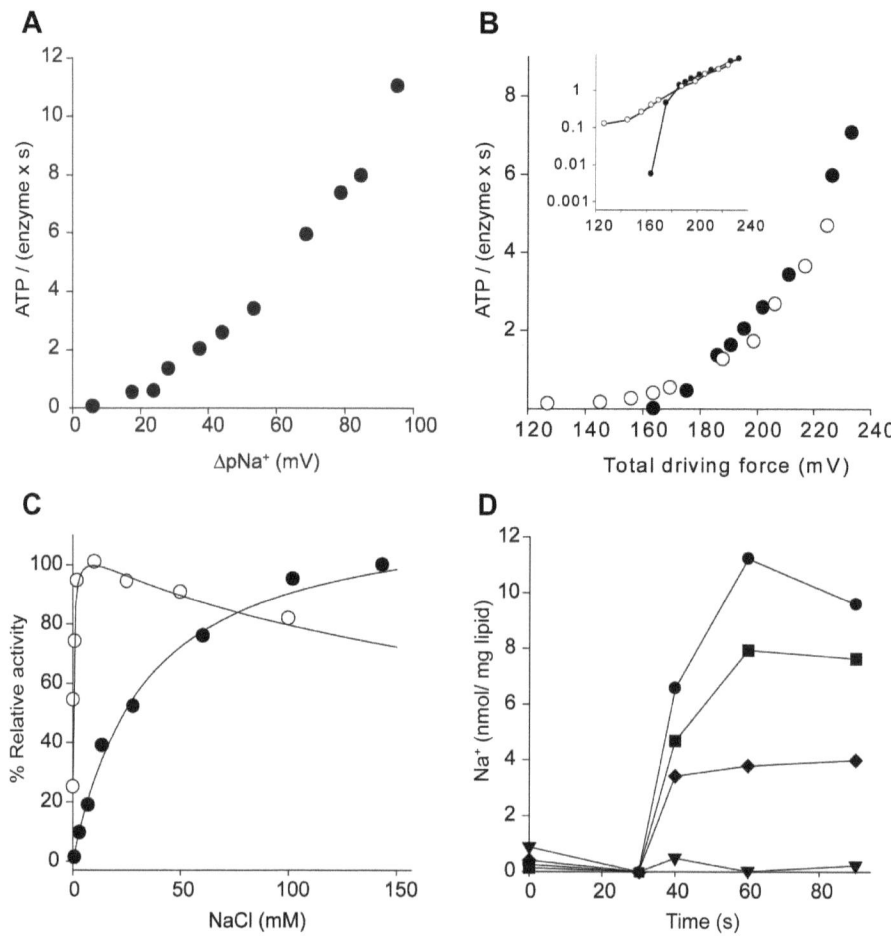

for the ion.

How can the same binding site display such different affinities for the coupling ion? Importantly, an essential arginine in subunit a, also known as the stator charge, has been shown to be indispensable for ion release, suggesting a role as potent modulator of the affinity (Elston et al. 1998; Wehrle et al. 2002). Between the two access pathways of the ion, the arginine is thought to form a complex with an empty ion binding site on the c-ring. In synthesis or hydrolysis direction, this complex is broken by replacement of the arginine by an incoming coupling ion from the P- or N-side, respectively. Different chemical environments of these access channels might account for the different affinities in ion binding.

According to these and earlier results, efficient ATP synthesis by the Na^+ or H^+ ATP synthase of $P.$ $modestum$ or $E.$ $coli$, respectively, requires contributions from the ion gradient and the electric potential. These forces, however, drive ATP synthesis only in support with a sufficiently high ion concentration at the P-side ($K_D \sim 35$ mM and pK_a 6.2 for the $P.$ $modestum$ and $E.$ $coli$ enzyme, respectively). A remarkably similar pK_a of 6.1 has been calculated for an unspecified site of the ATP synthase of $R.$ $capsulatus$ (Feniouk et al. 2004). We further show, that the occupation of this binding site is not influenced by $\Delta\psi$ (Table 3.1) and it is therefore suggested, that the electrical potential exerts its effect mainly along the cytoplasmic ion pathway, which according to accessibility studies is more hydrophobic than the P-side access pathway (Angevine et al. 2003). The electrical field might therefore together with the arginine promote the release of

---

Figure 3.1 *(preceding page)*: **Requirements for ATP synthesis of the $P.$ $modestum$ enzyme.**
A. Impact of ΔpNa. A constant $\Delta\psi$ of 160 mV was applied by a K^+/ valinomycin diffusion potential. The ΔpNa was varied using different NaCl concentrations in the assay buffer while keeping a constant internal Na^+ concentration (100 mM). **B. Impact of $\Delta\psi$ and ΔpNa.** Proteoliposomes containing 100 mM Na^+ and 0.5, 10 or 25 mM K^+ were prepared. At a constant ΔpNa of 120 mV, the size of $\Delta\psi$ was adjusted using different KCl concentrations in the assay buffer(open circles). Data from panel A were taken to show the influence of ΔpNa (closed circles). Inset: The data are plotted with the y-axis in a logarithmic scale. **C. Impact of Na^+ concentration.** Proteoliposomes for synthesis measurements (filled circles) contained Na^+ concentrations varying from 0.8 mM to 200 mM. To maintain constant $\Delta\psi$ of 140 mV and ΔpNa of 108 mV, proteoliposomes were diluted 1:60 into buffer containing 50 µM Na^+. The data were fitted yielding an apparent K_D of 35 mM. Data for the hydrolysis reaction (open circles) were taken from von Ballmoos and Dimroth (2007). **D. Na^+-transport through isolated F_0.** Unidirectionally oriented F_0 proteoliposomes containing 0.5 mM or 200 mM KCl were incubated with 100 mM ^{22}NaCl as described under Methods. To create a ΔpNa alone (filled squares), liposomes containing 200 mM KCl were diluted 1:50 into buffer containing 60 µM NaCl and 200 mM KCl. For measurements with a $\Delta\psi$ only (filled diamonds) and a combination of $\Delta\psi$ and ΔpNa (filled circles), liposomes containing 0.5 mM KCl were diluted 1:50 into buffer containing 100 mM or 60 µM NaCl, respectively. Inhibition of the ATP synthase by DCCD (filled triangles) was performed prior to the incubation with ^{22}NaCl to prevent occupation of the inhibition site by Na^+.

3 Glimpse into an ATP synthase's F_0 motor at work

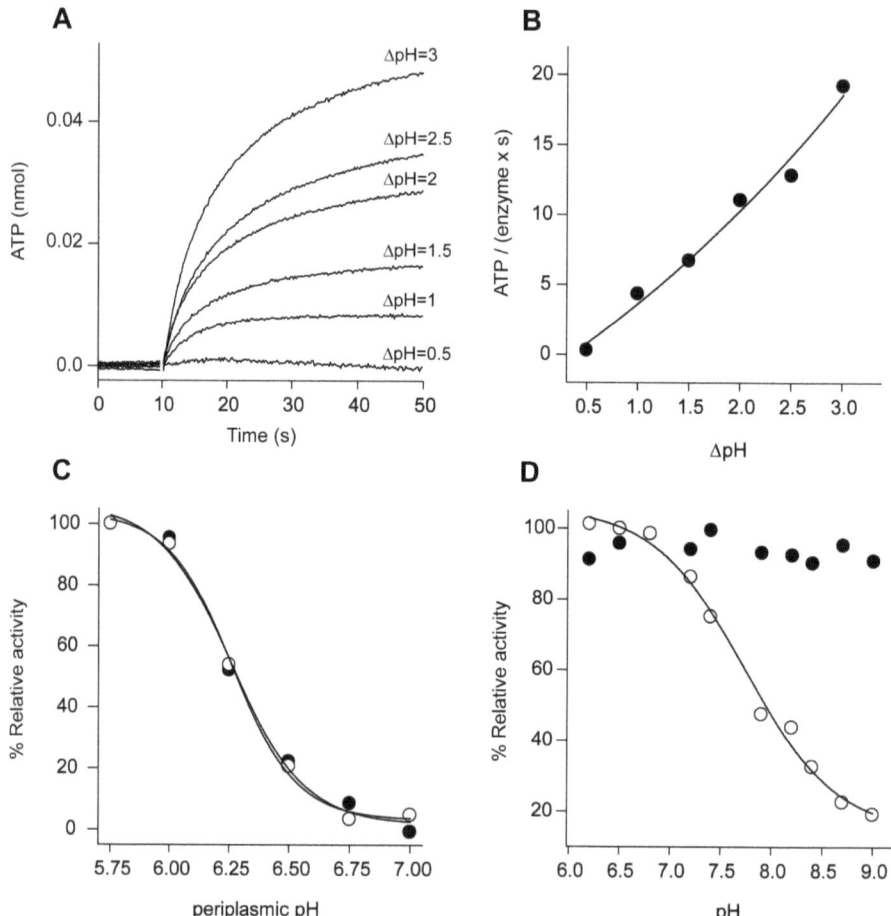

3.2 Glimpse into an ATP synthase's F_0 motor at work

the tightly bound ion from an incoming rotor site through the N-side access pathway. A general model for directed rotation during ATP synthesis is suggested in which the experimentally observed critical parameters for unloading and loading of the rotor sites at the appropriate location are at the core. (Fig. 3.4, see also Supplementary Discussion on page 58).

The present results affect the mechanism of ATP synthesis *in vivo*. The driving forces $\Delta\psi$ and ΔpNa in *P. modestum*, or $\Delta\psi$ and ΔpH in *E. coli* are created by the methylmalonyl-CoA decarboxylase Na^+ pump, or respiratory proton pumps, respectively. The high Na^+ concentration requirement for ATP synthesis in *P. modestum* is readily realized by its natural marine environment. For *E. coli*, however, which grows optimally at pH 7 - 8, but requires pH $<$ 6.5 for ATP synthesis *in vitro*, the underlying mechanism is difficult to reconcile. An obvious solution to this dilemma would be a local increase of the proton concentration at the ATP synthase, as proposed in the long standing hypothesis of lateral diffusion of the pumped protons along the membrane surface (Cherepanov et al. 2003; Williams 1978). The hypothesis is in accordance with the observation that low pH values ($<$ 6.5) are only required if ATP synthesis is energized by an artificial K^+/ valinomycin diffusion potential, whereas *E. coli* vesicles readily synthesize ATP at pH 7-8 if they are energized by respiratory proton pumping. Similar differences but with a shift to more alkaline pH values had previously been found for the different modes of energizing ATP synthesis in the alkaliphilic *Bacillus firmus* (Guffanti et al. 1984). ATP synthesis experiments with *Halobacterium salinarium* (Michel and Oesterhelt 1980a) and spectroscopic measurements with fluorophores at the membrane surface further support the idea of lateral proton diffusion (Branden et al. 2006; Heberle et al. 1994). To elucidate details of the ATP synthesizing machinery, we followed ATP synthesis, NADH oxidation, membrane potential and intravesicular pH in inverted *E. coli* membrane vesicles in

---

Figure 3.2 *(preceding page)*: Requirements for ATP synthesis of the *E. coli* enzyme.
A. Acid-base induced ATP synthesis. Twenty µl proteoliposomes were mixed with 20 µl Acid buffer (100 mM MES, pH 5.75) and incubated for 5 minutes at room temperature. Meanwhile, 450 µl Base-buffer (100 mM MOPS-NaOH, pH 6.25 - 7.25, or 100 mM Tricine-NaOH, pH 7.5 - 9, 150 mM KCl, 2 mM K_2PO_4, 2.5 mM $MgCl_2$) were mixed with 50 µl luciferase reagent, 0.75 mM ADP, 130 nM valinomycin and a baseline was recorded. ATP synthesis was initiated by rapid addition of the proteoliposomes into the reaction buffer. Indicated are the different ΔpH values. **B. Impact of pH on ATP synthesis rate.** Initial rates from the traces from figure 3.2 A were calculated. **C. Impact of pH at the *P*-side on ATP synthesis.** ATP synthesis measurements were performed as described in panel A. The pH was kept constant using Acid (100 mM MES, pH 5.75 - 7) and Base buffers (100 mM Tricine, pH 7.25 - 9) at different pH values. Shown are results for ΔpH = 1.5 (open circles) and ΔpH = 2 (closed circles). The two series were normalized to the values obtained at pH 5.75. **D. pH dependency of H^+-transport through the F_0 part.** Transport measurements through unidirectionally oriented F_0 parts were determined as described (Wiedenmann et al. 2008). Shown are the dependencies of transport from the N-side to the P-side (hydrolysis direction, closed circles) and the reverse direction (synthesis direction, open circles).

3 Glimpse into an ATP synthase's F_0 motor at work

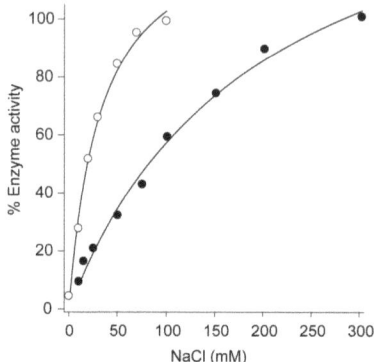

Figure 3.3: Impact of Na^+ concentration on ATP synthesis and ATP hydrolysis in the *I. tartaricus* cY70F mutant. The experiment was performed as described in figure 3.1 C, except that heterologously expressed enzyme of *Ilyobacter tartaricus*, containing mutation Y70F in the c-subunit, was used (closed circles). The data were fitted yielding an apparent $K_D > 190$ mM. Data for the hydrolysis reaction (open circles, $K_D = 33$ mM) were taken from von Ballmoos and Dimroth (2007).

parallel (Fig. 3.6 on page 61). The results show an initial phase of < 10 s, where NADH oxidation starts, the membrane potential is fully established, and ATP synthesis increases continuously. Importantly, only a marginal drop of the pH is observed (from 7.25 to 7.15) which is insufficient to meet the H^+ concentration requirement for ATP synthesis *in vitro*. We therefore intuitively suggest a significant pH drop by proton pumping at the membrane surface that is sensed by vicinal pyranine molecules and attenuation of the signal by the large excess of the pyranine molecules in the bulk which are not affected. We envisage that by modifying the system, pH changes at the membrane surface can be measured directly and will lead to detailed insights into the mechanism of lateral proton diffusion.

3.2 Glimpse into an ATP synthase's F_0 motor at work

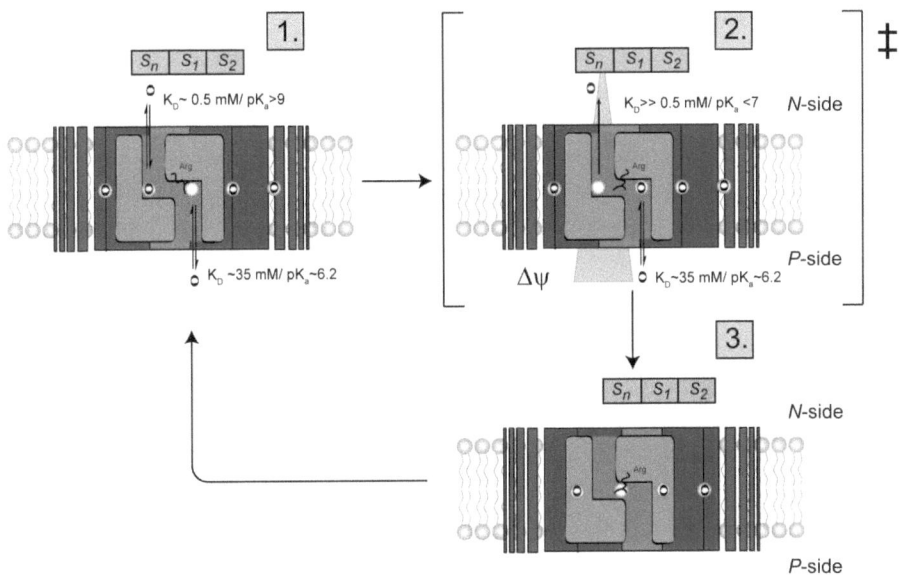

Figure 3.4: Cartoon illustrating events in the rotor/ stator interface during ATP synthesis. (rotation from left to right) Side view of the a/ c interface of an F_0 motor. Shown is the c-ring (dark grey) with binding sites (white circles) and subunit a containing the stator charge (Arg) and the P-side and the N-side access channel. ‡ indicates the transition state. **1.)** Between the opposing access pathways, the stator charge is connected to an unoccupied binding site (S_1) and Na^+ ions are bound to all other c-ring sites. In order to elicit rotation, the arginine must be released from S_1 and must form a new complex with the next incoming binding site (S_n) which requires binding of a coupling ion to S_1 and dissociation from S_n, where S_1 defines the first and S_n the last site of the rotating c-ring **2.)** At sufficiently high P-side coupling ion concentrations, an incoming ion replaces the arginine in the complex with S_1 which disconnects rotor and stator and allows S_1 with its neutralized charge to proceed into the interface with the lipids (Junge et al. 1997). In order to do so, S_n releases its ion into the N-side channel and forms a new complex with the stator charge, a process which is aided by the membrane potential. **3.)** The switch of the arginine between S_1 and S_n generates torque and completes one step of c-ring rotation and ion transport. Hence, ion concentrations on both sides of the membrane and the membrane potential affect the dissociation/ association equilibria of binding site/ arginine and binding site/ ion complexes at the two opposing access routes to cause directed rotation and torque generation under proper conditions as specified below.

$$S_1(Arg) + S_n(H^+/Na^+) + H^+/Na^+_{out} \rightleftharpoons S_1(H^+/Na^+) + S_n(Arg) + H^+/Na^+_{in} \quad (3.1)$$

3.3 Methods summary

Preparative procedures. ATP synthases from *P. modestum* and *E. coli* were purified as described (Neumann et al. 1998; Wiedenmann et al. 2008). Heterologously expressed ATP synthase from *Ilyobacter tartaricus* containing mutation Y70F in the c-subunit was purified as described (Vorburger et al. 2008). Inverted vesicles of *E. coli* cells were prepared by French Press disrupture in the presence of 100 mM buffer, 0.5 % Na^+-cholate and 1 mM pyranine. Subsequent centrifugation and gelfiltration were applied to remove excess Na^+-cholate and pyranine.

Assays. Na^+-transport and ATP synthesis measurements with the closely related enzymes of *P. modestum* and *I. tartaricus* were performed as described (Kluge and Dimroth 1992). H^+-transport measurements through unidirectionally reconstituted F_0 parts were performed as described recently (Wiedenmann et al. 2008). ATP synthesis measurements with the *E. coli* ATP synthase were done as described (Fischer and Gräber 1999) with modifications (see supplementary methods). Spectroscopic measurement of membrane potential generation was performed as described (Dröse et al. 2005).

Apparent K_D values (using a model for a single ligand binding) and initial transport rates (as described in Wiedenmann et al. (2008)) were calculated from original data with mathematical models using the program SigmaPlot (SyStat Software, Erkrath, Germany).

3.4 Supplementary Discussion

3.4.1 Asymmetric Na^+ binding affinities of the F_0 ATP synthase of *P. modestum*

At pH values around 6.5 in the absence of Na^+ ions, the F_0 part of the *P. modestum* ATP synthase translocates protons at a reduced rate (Laubinger and Dimroth 1989). This transport is abolished by Na^+, which blocks proton binding to the c-ring sites. We exploited this property to estimate the binding affinity for Na^+ from either side of the membrane. We prepared unidirectionally oriented F_0 liposomes from *P. modestum* and determined the initial rate of proton transport in either direction in the presence of varying Na^+ concentrations. In accord with the data for the holoenzyme, we determined apparent Na^+ binding affinities of ~ 1 mM and ~ 15 mM in hydrolysis and synthesis direction, respectively (Fig. 3.5).

3.4 Supplementary Discussion

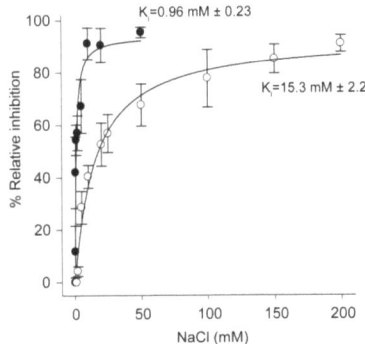

Figure 3.5: Inhibition of $\Delta\psi$-driven unidirectional H^+-transport through *P. modestum* F_0 ATP synthase by Na^+ ions. Liposomes containing unidirectionally reconstituted F_0 parts were prepared as described (Wiedenmann et al. 2008) and incubated overnight with Na^+ concentrations varying from 50 µM to 200 mM in the presence of either 0.5 or 50 mM K_2SO_4. For H^+-transport measurements in hydrolysis direction, 10 µl of liposomes containing 50 mM K_2SO_4 were mixed with 2.5 ml buffer (2 mM MOPS-Tris, pH 6.5, 0.5 mM K_2SO_4) of the same Na^+ concentration and the transport was initiated by the addition of 16 nM valinomycin. For H^+-transport measurements in synthesis direction, 10 µl of liposomes containing 0.5 mM K_2SO_4 were mixed with 2.5 ml buffer (2 mM MOPS-Tris, pH 6.5, 50 mM K_2SO_4) of the same Na^+ concentration and the transport was initiated by the addition of 16 nM valinomycin. Initial slopes were determined as described (Wiedenmann et al. 2008). Average values and standard deviations from three measurements of two independent experiments are shown.

3.4.2 Spectroscopic surveillance of critical parameters during ATP synthesis in inverted membrane vesicles of *E. coli*

It has been proposed, that protons, which are pumped out of the cell can diffuse laterally along the membrane surface before they equilibrate with the bulk (Cherepanov et al. 2003). In such a scenario, primary pumps like the respiratory chain complexes could acidify the membrane surface and lower the local pH around the ATP synthase sufficiently to allow ATP synthesis at optimal rates. We mimicked a minimal cellular system by preparation of inverted membrane vesicles of *E. coli* cells. The vesicles were energized by addition of NADH (or succinate) and critical parameters (NADH oxidation, ATP synthesis, creation of a membrane potential, intravesicular pH) were followed (Fig. 3.6, panels 1-4). As expected, after NADH oxidation had ceased, ATP synthesis stopped and the electrical potential was abolished. Additionally, a rapid pH decrease was observed with pyranine, which was almost fully reversed after the NADH had been consumed. However, the pH drop was far too small (from pH 7.25 to pH 7.15) to establish the proton concentration requirement for ATP synthesis. This discrepancy can be accounted

3 Glimpse into an ATP synthase's F_0 motor at work

for if the pH decrease by the pumped protons is essentially restricted to the membrane surface. As a consequence, the pH decrease will only be sensed by pyranine molecules in the proximity of the membrane surface and the signal will be attenuated by excess pyranine molecules in the bulk, which are not affected. Further inspection of the results indicated that in an initial phase of < 10 s, the membrane potential was fully established and 80 % of the pH decrease took place, while ATP synthesis was still suboptimal. After this phase, the internal pH remained almost constant and the ATP synthesis rate became maximal. Taking these data together, we propose that the pumped protons diffuse along the membrane acidifying the environment of the ATP synthase, so that ATP synthesis is initiated. Similar results were obtained, when membrane vesicles were prepared at pH 8 (pH drop from 8 to $\sim$ 7.8). Data at pH 7.25 are shown, because the response of pyranine is maximal in this pH range.

3.4.3 Molecular mechanism of the F_0-ATP synthase in different operation modes

The model in the main text depicts critical events in the subunit a/ c interface during ATP synthesis (Fig. 3.4 on page 57). However, the ATP synthase can also work in reverse as an ATP-driven ion pump. Additional knowledge of the F_0 motor operation stems from transport experiments through the isolated F_0 part. We therefore probed our new model with the different modes of operations and find it compatible with all experimental findings.

I. Na_{in}^+/ Na_{out}^+ -exchange in the absence of any driving force. In the absence of any driving force, the enzyme of *P. modestum* catalyzes a 1:1 exchange of Na^+-ions between the two sides of the membrane (Kluge and Dimroth 1992). Our model suggests that the complex between the empty binding site (S_1) and the stator charge is not broken during this process. Accordingly, the exchange reaction is catalyzed by the neighboring binding site (S_2) in between the *P*-side access pathway and the c-ring/ lipid interface. Exchange via this route is possible because Na^+ ions on the binding sites in the c-ring/ lipid interface undergo direct exchange with Na^+ in the cytoplasm (Meier et al. 2003; Murata et al. 2000; von Ballmoos et al. 2002b).

II. ATP-driven ion pumping. When the F_1F_0 ATP synthase works in reverse as an ion pump, the rotational torque is generated by the hydrolysis of ATP in the F_1 part. In contrary to ATP synthesis, this reaction also occurs in solution in the absence of transmembrane forces. We envision that during ATP hydrolysis, an occupied binding site enters into the *P*-side access route, where the ion is released by electrostatic repulsion of the stator charge (Wehrle et al. 2002). Due to the low affinity at this location ($K_D(Na^+) > 35$ mM/ $pK_a < 6.2$), ion release

3.4 Supplementary Discussion

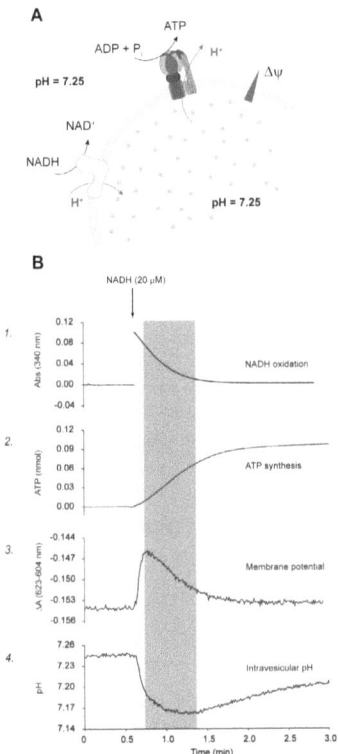

Figure 3.6: Time resolved monitoring of ATP synthesis in inverted vesicles of *E. coli*. A. Cartoon scheme of inverted vesicles. Shown are the events, which were followed spectrophotometrically in figure 3.6 B). Stars indicate the homogenous distribution of the pH indicator pyranine within the vesicles. **B.** Time traces of critical parameters during ATP synthesis. Shown are traces of NADH oxidation (**1**), ATP synthesis (**2**), buildup of a membrane potential (**3**) and the intravesicular pH (**4**). The reaction was initiated by addition of 20 µM NADH. Indicated are an initial phase (0 - 8 s, light grey) and a late phase (8 - 45 s, dark grey).

is facile, although inhibition has been observed at concentrations well above 150 mM Na^+ in the *I. tartaricus* enzyme (Xing et al. 2004). The newly formed complex between the empty binding site and the stator charge is not broken by an ion gradient or an electric potential as in ATP synthesis but by the mechanic force imposed by the F_1 motor. Once the site has been moved into the N-side access pathway, it is easily occupied due to the high binding affinity at this position and rotation is guaranteed.

3 Glimpse into an ATP synthase's F_0 motor at work

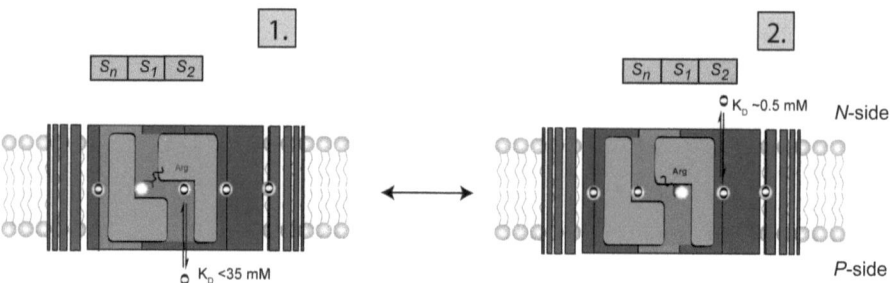

Figure 3.7: Cartoon showing events during Na^+_{in}/Na^+_{out}-exchange. Side view of the a/ c interface of a F_0 motor. Shown is the c-ring (dark grey) with binding sites (white circles) and subunit a containing the stator charge (Arg) and the P-side and the N-side access channel. In the absence of any driving force, the interaction between the stator charge (Arg) and binding site S_1 remains intact. The Na^+_{in}/Na^+_{out}-exchange is catalyzed by binding site S_2, which can shuttle between the periplasmic access channel (**1**) and the c-ring/ lipid phase interface (**2**).

III. Ion transport through F_0 in synthesis direction. The mechanism for ion transport through F_0 in synthesis direction is in principle identical to that described in figure 3.4 on page 57. However, the different apparent proton binding affinities for ATP synthesis and H^+-translocation in the *E. coli* enzyme (pK_a values of 6.2 and 7.7, respectively) display a certain mismatch. We speculate, that the presence of F_1 (higher load and less mobility of $\gamma\epsilon$ in the $\alpha_3\beta_3$ headpiece due to intersubunit contacts in the nucleotide binding sites (Feniouk et al. 2005)) hampers the free diffusional motion of the c-ring versus the stator. Higher coupling ion concentrations are therefore required to efficiently displace the arginine from the binding site and initiate a cycle of ion transport. A similar mismatch in affinities can also be deduced from the *P. modestum* data (35 mM for ATP synthesis and 15 mM for ion translocation).

It is noteworthy, that ATP synthesis in *P. modestum* or *E. coli* is driven by a ΔpNa or ΔpH alone at very low rates ($\sim$ 1-5 % of maximal rate). In these experiments however, the ion concentrations on the P-/ N-side (100 mM/ 0.5 mM Na^+ in *P. modestum* and pH 4.7/ 8.8 in *E. coli*) are close to the respective binding site affinities in the different access pathways. With $\Delta\psi$ alone ATP is not synthesized but ion transport through F_0 is observed, indicating a similar difference in motional freedom as speculated above. In summary, one may conclude, that the two crucial events of loading and unloading of the binding site from the P-side and to the N-side, respectively, are rather parallel than serial processes, as depicted in equation 3.1 on page 57. The probability of these two events happening together would then be an important determinant of enzyme catalysis (ATP synthesis or ion transport), which might be influenced by the freedom to move the c-ring against the stator (see also Feniouk et al. (2004)).

IV. Ion transport through F_0 in hydrolysis direction. For the *E. coli* enzyme, H^+-transport in hydrolysis direction resembles that in synthesis direction with opposite prefixes. The situation is more complex for the Na^+-translocating enzyme from *P. modestum*, where Na^+-transport in hydrolysis direction is driven by $\Delta\psi$ but not by ΔpNa, whereas in the opposite direction, ΔpNa (or $\Delta\psi$) is a potent driving force. ΔpNa-driven Na^+-transport (in hydrolysis direction) did not even occur at *P*-side Na^+-concentrations < 100 μM, where with a K_D of ~ 35 mM ion release should not be rate-limiting. We therefore conclude that occupation of the *N*-side binding site does not occur in the absence of a $\Delta\psi$ in spite of large coupling ion concentrations, thus reinforcing the interpretation, that $\Delta\psi$ acts mainly along the *N*-side access channel. In this transport direction, the $\Delta\psi$ has the opposite sign as in ATP synthesis and the ion is therefore pushed into the binding site. Whether the $\Delta\psi$ mainly targets the Na^+ ion or weakens the arginine/ binding site complex, can not be deduced from our data (see also ATP hydrolysis driven ion transport). Similar differences in driving ion transport through F_0 in the opposing directions are not observed for the *E. coli* enzyme, where $\Delta\psi$ and ΔpH are equivalent driving forces (Wiedenmann et al. 2008). The differences between the two enzymes may result from different stabilities of the arginine-binding site complex, as discussed (Vorburger et al. 2008).

3.4.4 Final conclusions

Here, we define the requirements for ATP synthesis and ion translocation through the isolated F_0 part of a H^+ and a Na^+ translocating ATP synthase in both working directions. Of crucial importance is the switch in the affinity of the binding site as it moves from one to the opposing ion access pathway. The enzyme's function therefore depends on the conditions for loading and unloading of the binding site in each specific access pathway, i.e. the ion gradient (or the effective coupling ion concentrations on either side) and the electrical potential. In the present model of the molecular mechanism the stator charge arginine has a key function in the generation of torque. Under proper conditions for loading and unloading of the binding site, an incoming coupling ion displaces the arginine from its complex with an empty binding site, allowing it to form a new complex with an neighbouring binding site from which the ion gets displaced, and so on. The necessity of the binding site to strip off its ion in order to move with the arginine between the opposing access routes effectively prevents ion leakage (Elston et al. 1998). If the arginine is replaced by an uncharged amino acid, however, a site may shuttle between the opposing access pathways in the empty or ion bound state. This transforms the a/ c interface into a non-rotating transporter as observed experimentally (data not shown and Valiyaveetil and Fillingame (1997); Wehrle et al. (2002)).

3 Glimpse into an ATP synthase's F_0 motor at work

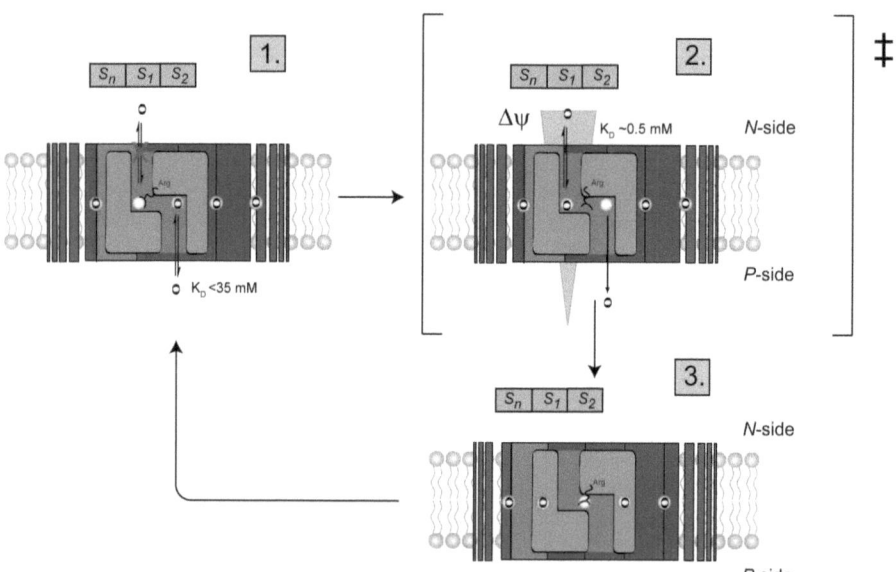

Figure 3.8: Cartoon showing events during $\Delta\psi$-driven Na^+ translocation of *P. modestum* F_0 in hydrolysis direction (rotation from right to left). Side view of the a/ c interface of a F_0 motor. Shown is the c-ring (dark grey) with binding sites (white circles) and subunit a containing the stator charge (Arg) and the P-side and the N-side access channel. ‡ indicates the transition state.
1.) **In the absence** of any driving force, the $S_1(Arg)$ complex can not be broken by Na^+ uptake through the N-side channel, irrespective of the coupling ion concentration on this side. **2.)** If an N-side positive is applied, the $S_1(Arg)$ complex is disrupted forming a transition state where ions entering through the N-side channel form the $S_1(Na^+)$ complex and the positively charged arginine enforces the dissociation of the Na^+ form the $S_2(Na^+)$ into the P-side channel. **3.)** By forming the $S_2(Arg)$ complex, the rotor has moved one step into the opposite direction as in the ATP synthesis direction.

3.5 Supplementary Methods

3.5.1 Enzyme preparation and reconstitution into phospholipid vesicles

Expression, purification and reconstitution into phospholipid vesicles of ATP synthases from
P. modestum, *I. tartaricus* and *E. coli* were done by described methods (Laubinger and Dimroth 1988; Neumann et al. 1998; Wiedenmann et al. 2008). Preparation of phospholipid vesicles containing F_0 parts of the enzymes of *P. modestum* and *E. coli* was done as described (Kluge and Dimroth 1992; Wiedenmann et al. 2008). The unidirectional orientation of the enzymes of *P. modestum* (> 90 % N-side out) and *E. coli* (> 95 % N-side out) was confirmed with ATP hydrolysis measurements and Western-Blotting as described (Wiedenmann et al. 2008). The amount of incorporated enzyme was confirmed with ATP hydrolysis measurements and Western-Blotting as described (Wiedenmann et al. 2008).

3.5.2 Na^+-transport measurements

Highly concentrated F_0 liposomes (400 mg lipid/ ml) were equilibrated overnight in the appropriate buffer containing 5 µCi/ ml ^{22}NaCl (GE Healthcare, Glattbrugg, Switzerland). Pre-equilibrated liposomes were diluted into the appropriate assay buffer to a final lipid concentration of 8 mg/ ml. The transport reaction was initiated by addition of 1 µM valinomycin.

3.5.3 H^+-transport measurements

H^+-transport with the F_0 part of the *E. coli* ATP synthase were preformed as described (Wiedenmann et al. 2008).

3.5.4 ATP synthesis measurements with *P. modestum* liposomes

ATP synthesis was monitored using a continuous luciferase assay. To 1 ml assay buffer (15 mM Tris-phosphate, pH 7.5, 2.5 mM $MgCl_2$, and different amounts of KCl and/ or NaCl, as indicated) 0.7 mg proteoliposomes, 0.5 mM ADP and 10 µl Bioluminescence assay CLS II (Roche Diagnostics, Rotkreuz, Switzerland) were mixed and the reaction started with 1 µM valinomycin. The luminescence was recorded with a rate of 5 Hz in a luminometer (Turner Biosystems, Sunnyvale, USA). Luminescence signals were quantified by addition of 0.2 nmol ATP at the end of each time trace. Initial rates were calculated using the slope of the first three seconds, omitting the signal change after injection.

3.5.5 Reconstitution and ATP synthesis measurements with *E. coli* ATP synthase

The purified ATP synthase of *E. coli* was reconstituted with modifications as described (Wiedenmann et al. 2008). Briefly, soybean phosphatidylcholine (Type II, Sigma-Aldrich, Buchs, Switzerland) was dissolved at a concentration of 30 mg/ ml in buffer A (10 mM HEPES-NaOH, pH 7.5, 10 mM $MgSO_4$, 0.1 mM Na_2-EDTA, 100 g/ l sucrose) and sonicated at 7.5 µ for 2 x 30 s on ice using a tip sonicator (Sanyo MSE Soniprep, München, Germany) to form unilamellar liposomes. The suspension was adjusted to 1 % Na^+-cholate (from a 10 % stock solution containing phosphatidic acid (15 mg/ ml)) and mixed with F_1F_0 ATPase (lipid : protein ratio (w/ w) = 1:50). The mixture was incubated for 20 min at room temperature and subsequently 1 ml was loaded on a PD-10 column (GE Healthcare, Glattbrugg, Switzerland), preequilibrated with buffer A. Turbid fractions were pooled and the proteoliposomes collected by centrifugation for 45 min at 200,000 g at 4 °C and resuspended in buffer A (60 mg/ ml). ATP synthesis was monitored using the continuous luciferase assay. Twenty µl proteoliposomes were mixed with 20 µl Acid buffer (100 mM MES, pH 5.75-7) and incubated for 5 min at room temperature. Meanwhile, 450 µl Base-buffer (100 mM MOPS-NaOH, pH 6.25 - 7.25, or 100 mM Tricine-NaOH, pH 7.5 - 9, 150 mM KCl, 2 mM K_2PO_4, 2.5 mM $MgCl_2$) were mixed with 50 µl luciferase reagent, 0.75 mM ADP, 130 nM valinomycin and a baseline was recorded. The ATP synthesis reaction was started by rapid mixing of the acidified proteoliposomes with the Base-buffer. Initial rates were calculated using the slope of the first three seconds, omitting the signal change after mixing. ATP synthesis in inverted membrane vesicles *E. coli* DK8 cells were transformed with plasmid pBWU13 (Moriyama et al. 1991) and were cultivated as described (Wiedenmann et al. 2008). Cells were washed with 10 mM Tris-Cl, pH 8 and suspended in buffer S (100 mM MOPS-NaOH, pH 7.5, 50 mM NaCl, 4 mM Na_2PO_4, 2 mM $MgSO_4$), supplemented with 0.5 % Na-cholate and 1 mM pyranine. Inverted membrane vesicles were prepared by French-Press and ultracentrifugation as described (Moriyama et al. 1991). Remaining Na^+-cholate and pyranine was removed by gelfiltration on a PD-10 column, preequilibrated with buffer S. The turbid fractions were pooled and directly used for ATP synthesis measurements as described (Etzold et al. 1997).

Acknowledgements: We thank Martin Badertscher and Klaus Girgenrath (Laboratory of Organic Chemistry, ETH Zürich) for help with γ-counting.

4 Impact of ΔpNa on the F_0 part from *Propionigenium modestum*

4.1 Introduction

F_1F_0 ATP synthases are responsible for the production of the majority of ATP, the universal energy currency in living cells. These enzymes synthesize ATP from ADP and inorganic phosphate by a rotary mechanism utilizing the electrochemical gradient provided by oxidative phosphorylation, decarboxylation phosphorylation or photophosphorylation (von Ballmoos et al. 2008). The vast majority of F-type ATPases use protons as their coupling ions, but those of some anaerobic bacteria use Na^+ ions instead.

The structure of the enzyme is bipartite with a soluble F_1 part and a membrane embedded F_0 part. In bacteria, the catalytic F_1 part consists of subunits $\alpha_3\beta_3\gamma\epsilon\delta$ and is connected to the F_0 domain via a peripheral stalk which is built up by a b_2-dimer and a central stalk which is constituted by the $\gamma\epsilon$-complex. The F_0 part in bacteria contains one a-subunit, the b_2-dimer and 10-15 c-subunits which assemble into the c-ring.

During ATP synthesis, the flux of H^+ or Na^+ through F_0 following the electrochemical potential is used to drive the rotation of the c-ring versus the stator subunits $ab_2\alpha_3\beta_3\delta$. This torque is applied to the central stalk which drives conformational changes in the catalytic F_1 part that ultimately lead to synthesis of ATP (for review, see Capaldi and Aggeler (2002); Dimroth et al. (2006)).

During ATP synthesis, it is envisaged that coupling ions enter the F_0 part from the periplasm through an aqueous pathway located within subunit a and are bound to the binding sites on the c-ring. From there, they are released into the cytoplasmic reservoir through a poorly understood pathway (Dimroth et al. 2006). We recently showed that the F_1F_0 ATP synthase of *P. modestum* requires an ion concentration gradient (ΔpNa) of at least 30 mV to synthesize ATP. This finding was unexpected since Na^+-transport through the isolated F_0 part was only possible in presence of a membrane potential (Kluge and Dimroth 1992). In these experiments an ion concentration gradient was without effect on Na^+-transport (Kluge and Dimroth 1992). No such differences are observed in the H^+-dependent F_0 part of the E. coli ATP synthase where

4 Impact of ΔpNa on the F_0 part from Propionigenium modestum

$\Delta\psi$ and ΔpH serve as equivalent driving forces for both transport directions (Wiedenmann et al. 2008).

In order to shed light on this apparent discrepancy we investigated the ion translocation pathway of the F_0 part of the *P. modestum* in more detail. We first showed that the orientation of the F_0 parts in our liposome preparations is predominantly unidirectional with the cytoplasmic side of the enzyme facing outwards. This finding allowed us to test whether different properties of the ion translocation pathway during transport through isolated F_0 in hydrolysis and synthesis direction are responsible for different effects of ΔpNa.

4.2 Material and Methods

4.2.1 Enzyme purification and reconstitution

ATP synthase from *Propionigenium modestum* was isolated and purified as described (Laubinger and Dimroth 1988). To obtain F_0-containing liposomes, the ATP synthase was reconstituted and F_1 parts were removed using a low salt buffer at high pH as described (Kluge and Dimroth 1992; Laubinger and Dimroth 1988). Liposomes for ATP synthesis experiments were prepared as described (Dmitriev et al. 1993) with modifications. Briefly, phosphatidylcholine (Sigma-Adrich, Steinheim, Germany) was dissolved at a concentration of 40 mg/ml in 15 mM Tris-PO_4 pH 7.5, 200 mM NaCl and 0.25 mM K_2-EDTA. The suspension was sonicated for 60 s with 7.5 μ amplitude in a tip sonicator (MSE Scientific Instruments, Crawley, UK) to form unilamellar liposomes and adjusted to 2.5 mM $MgCl_2$. Purified ATP synthase from *P. modestum* was added in a protein to lipid ratio of 1:80 (w/w). The mixture was incubated for 20 min on ice, frozen in liquid nitrogen for 5 min and thawed in a waterbath at 4 °C. Proteoliposomes were subsequently collected by centrifugation for 45 min at 200,000 g and resuspended to a final lipid concentration of approximately 150 mg/ml.

4.2.2 Measurement of ATP hydrolysis activity

ATP hydrolysis measurements were performed using the coupled enzyme assay as described (Laubinger and Dimroth 1988) with the following modifications. Instead of potassium phosphate, 50 mM Tris-Cl, pH 7.5 was used and Triton X-100 was omitted in the experiments with intact proteoliposomes as indicated.

4.2.3 Western Blot analysis

Western blots using antibody against subunit β (anti-rabbit, ABCam, Cambridge, UK) were preformed according to the protocol of the manufacturer. Secondary antibody incubation was performed with horseradish peroxidase conjugated to anti-rabbit antibody and protein bands were visualized using the ECL detection system (GE Healthcare, Glattbrugg, Switzerland). The relative amounts of protein were quantified after scanning and analyzing the blots using the QuantityOne software (BioRad, Hercules, USA).

4.2.4 Na^+-transport measurements

Na^+-influx was determined using $^{22}Na^+$ as described previously (Kluge and Dimroth 1992). For measurement of Na^+-efflux, a highly concentrated liposome suspension (400 mg/ml) was incubated overnight with 100 mM NaCl (10 µCi ^{22}NaCl) to allow equilibration with the internal liposomes volume. Radioactivity was quantified using a γ-counter (Cobra II, Canberra Packard, Schwadorf, Austria).

4.2.5 Determination of ATP synthesis

The ATP synthesis reaction was started by rapid mixing of 300 µl injection buffer with 300 µl assay buffer inside the luminometer (Glomax, Turner Biosystems, Sunnyvale, USA). Luminescence signal was recorded at a rate of 5 Hz. The injection buffer contained 10 mM Tris-PO$_4$ pH 7.5, 2.5 mM MgCl$_2$, 0.5 mM ADP and 1 µM valinomycin. The assay buffer contained 15 mM Tris-PO$_4$, 2.5 mM MgCl$_2$, 0.5 mg liposomes and 30 µl luciferase assay (ATP Bioluminescence Kit CLS II, Roche Diagnostics, Mannheim, Germany). For quantification of synthesized ATP, 0.1 nmol ATP was added at the end of each experiment.

4.3 Results and Discussion

4.3.1 Orientation of *P. modestum* F_0 part in liposomes

The ATP synthase from *P. modestum* was reconstituted using the freeze/ thaw procedure as described (Laubinger and Dimroth 1988). The F_1 part was subsequently removed by incubation with a low-salt buffer, i.e. the orientation of the enzyme was not altered by this procedure (Kluge and Dimroth 1992). In the first series of experiments, we determined the efficiency of this reconstitution procedure and the orientation of the incorporated ATP synthases. The reconstitution efficiency was estimated by comparing ATP hydrolysis activities before and after removal of unincorporated ATP synthase by centrifugation yielding an approximate value of 53 %. The activities were measured in presence of Triton X-100 in order to make the membrane impermeable nucleotides available to inwards oriented F_1 headpieces.

Next, we incubated the liposomes in a low salt buffer to strip the external F_1 parts. Efficient removal of the F_1 part by centrifugation was confirmed by measuring ATP hydrolysis (in the absence of Triton X-100) which revealed that > 95 % of the external F_1 heads have been removed by this procedure. Finally, the orientation of the enzyme was calculated by determining activities in the presence of 0.1 % Triton X-100 before and after incubation with stripping buffer. This approach is slightly hampered by the fact that Triton X-100 stimulates hydrolysis activity of the ATP synthase which leads to an overestimation of the remaining F_1 parts (Laubinger and Dimroth 1987). Nevertheless, more than 91 % of hydrolysis activity was lost by removing external F_1. About 3 % of the residual activity can be assigned to remaining external F_1 and $\sim$ 5 % to internal F_1 parts. In conclusion, $\sim$ 95 % of the ATP synthases incorporated into liposomes face with their cytoplasmic side towards the external buffer.

In a second approach, the amount of subunit β in the liposome preparation was followed at different stages of the reconstitution procedure by Western Blotting. The reconstitution efficiency was estimated by comparing the signal before and after washing the liposomes. Values around 70 % were obtained which is modestly higher than the efficiency obtained by activity measurements. The reason for the apparent increase in reconstitution efficiency is likely to be due to detection of inactive enzyme. Similarly, the fraction of inside-oriented F_1 parts was slightly higher with approximately 7 %. However, both approaches independently confirmed a uniform orientation of > 90 % of ATP synthase complexes.

4 Impact of ΔpNa on the F_0 part from Propionigenium modestum

Table 4.1: Orientation of ATP synthase after reconstitution into liposomes.
ATP hydrolysis activities in presence of 0.1 % Triton X-100 directly after the freeze/ thaw procedure, after washing of proteoliposomes and after stripping the F_1 part are shown in the first column. The respective fractions are shown in the third column. The intensities and the corresponding fractions from Western blot analysis are shown in columns 2 and 4, respectively. For calculation of inside-oriented F_1 parts the values obtained after washing the liposomes were divided by the values obtained after removal of F_1, i.e. not incorporated enzyme was excluded from the calculation.

sample	activity (U)	intensity (AU)	% activity	% intensity
after freeze/ thaw	1.7	1307	100	100
after washing	0.9	903	53	69
after stripping	0.08	66	5	5

4.3.2 ΔpNa is an effective driving force in synthesis, but not hydrolysis direction

It was shown that Na^+-transport through isolated F_0 can not be energized by a Na^+ concentration gradient (ΔpNa) but requires a membrane potential (Δψ) instead (Kluge and Dimroth 1992). The finding that ATP synthesis is obligatorily depended on a ΔpNa > 30 mV was therefore unexpected and difficult to explain. Investigating the orientation of ATP synthase in the proteoliposomes revealed that under the experimental conditions applied in the previous study only Na^+-transport in hydrolysis direction was observed through the uniformly oriented F_0 parts in the proteoliposomes, which is the opposite transport direction as during ATP synthesis. For determination of Na^+-transport in synthesis direction, liposomes were incubated with high concentrations of radioactive $^{22}Na^+$ overnight to achieve equilibration with the internal liposome volume.

In their original work, Kluge and Dimroth have conducted similar experiments. In their study, they applied an outward rectified ΔpNa of 86 mV but observed only marginal efflux of Na^+. After 5 min more than 90 % of the radioactive Na^+ was still present inside the liposomes.

We modified their experiments by adding valinomycin to the reaction and including high K^+ concentrations within and outside the liposomes to prevent generation of a counteracting membrane potential through the electrogenic efflux of Na^+. As expected, only after addition of valinomycin, a rapid efflux of more than 80 % of radioactive Na^+ within 30 s was observed. In the following experiments the same liposomes were energized either by Δψ, ΔpNa or a combination of both driving forces. In all cases Na^+-transport was observed, however the highest transport rates were observed if Δψ and ΔpNa were applied concomitantly reinforcing the findings from ATP synthesis, that both forces are additive (see section 4.3.4). The determination of Na^+-efflux has several disadvantages compared to influx measurements: long incubation times

4.3 Results and Discussion

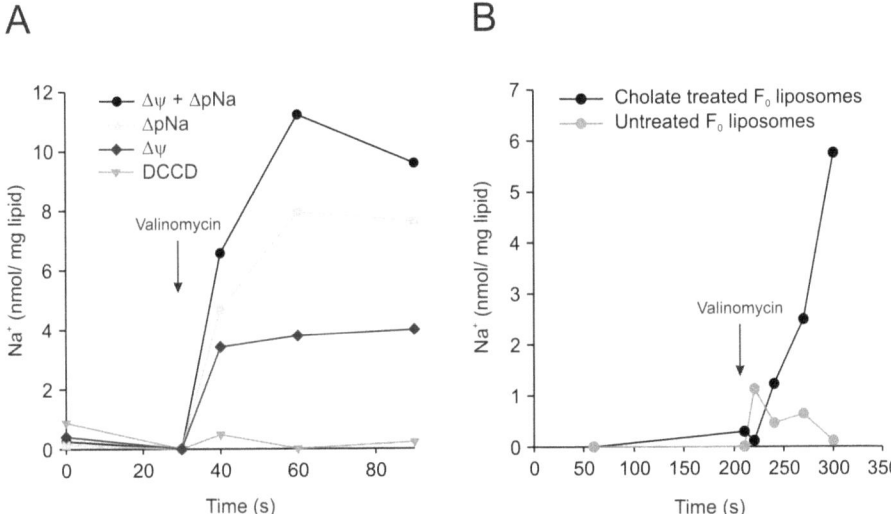

Figure 4.1: ΔpNa is an effective driving force for F_0 in synthesis direction.
A. Na^+-efflux with different driving forces is shown. Liposomes energized by ΔpNa contained 100 mM NaCl and 200mM KCl. The assay buffer contained 3 mM NaCl and 200 mM KCl. To apply both driving forces simultaneously, 1 mM KCl and 100 mM NaCl was present inside the liposomes. For Δψ-driven efflux, liposomes contained 100 mM NaCl and 1 mM KCl whereas 100 mM NaCl and 200 mM KCl were present in the assay buffer. **B.** Na^+-influx after cholate treatment. The liposome preparation contained 1 mM NaCl and 100 mM KCl. The external buffer contained 100 mM NaCl and 100 mM KCl. At the time point indicated, valinomycin was added to a final concentration of 1 μM. Control liposomes are identical, except they were not treated with cholate.

are required to ensure equilibration of Na^+ over the liposomal membrane, the efflux is very fast due to the small liposomal volume and large amounts of radioactive Na^+ are required. As a consequence, only one to two points are recorded before the reaction is finished. Therefore, a second series of experiment aimed at circumventing these drawbacks of the efflux reaction. We reasoned that solubilizing the liposomes with 1 % cholate would allow some F_0 parts to change their orientation and face with the periplasmic side outwards. After cholate treatment of F_0 containing liposomes, detergent was removed by passing the suspension over a gelfiltration column. The cholate treated liposomes should be capable of ΔpNa-driven Na^+-influx. In parallel, liposomes without cholate treatment served as control. As expected, ΔpNa-driven Na^+-transport was observed exclusively in liposomes with cholate treatment but not in the control liposomes, further supporting the notion that ΔpNa is an effective driving force in synthesis direction but not in hydrolysis direction.

4 Impact of ΔpNa on the F_0 part from Propionigenium modestum

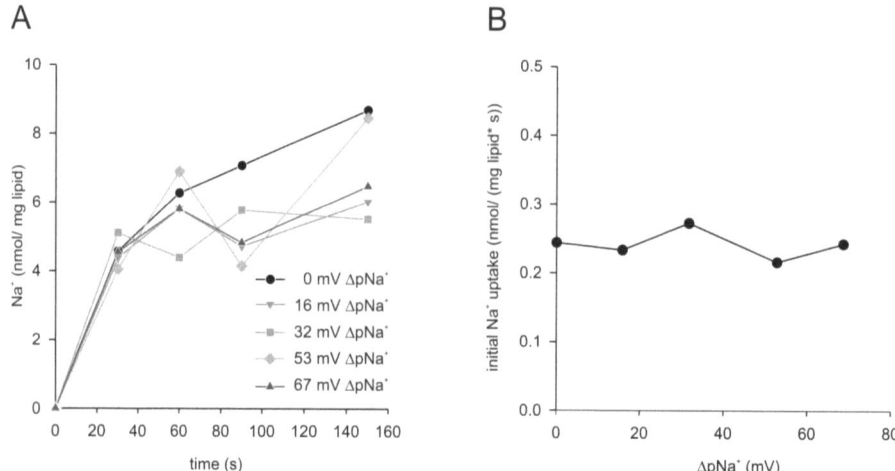

Figure 4.2: ΔpNa has no effect on Δψ-driven Na^+ uptake in hydrolysis direction.
A. Na^+ uptake was energized with a constant Δψ of approximately 80 mV while an inverse ΔpNa of different magnitudes was applied concomitantly. **B.** The initial rates are plotted versus the ΔpNa.

4.3.3 ΔpNa does not affect Δψ-driven Na^+-transport in hydrolysis direction

In experiments, where Na^+-transport in hydrolysis direction was driven by Δψ, a concomitant ΔpNa in the same direction was without effect on initial Na^+-transport rates. In consideration of these results, similar experiments with an opposing ΔpNa were preformed. Surprisingly, the initial rates were unaffected by ΔpNa. in the absence of any net driving force (ΔpNa = -Δψ), the initial rate remained unchanged (data not shown). These results suggest considerable differences in the mechanism of ion translocation in either direction. Apparently the presence of a Δψ which is positive on the N-side of the ATP synthase gears the F_0 part for transport in hydrolysis direction and eliminates influence of ΔpNa.

4.3.4 The K_{app} for Na^+ during ATP synthesis is much higher than during ATP hydrolysis

In addition to the different impact of ΔpNa during Na^+-transport through the F_0 part in hydrolysis or synthesis direction, considerable differences in the apparent Na^+ affinities were found during ATP synthesis and ATP hydrolysis The apparent affinity for Na^+ during ATP hydrolysis was determined to be in the range of 0.5 - 0.8 mM at pH 7 (Kluge and Dimroth 1993a;

4.3 Results and Discussion

Laubinger and Dimroth 1989). In contrast to this result, an apparent K_D of 35 mM for Na^+ was determined from initial ATP synthesis rates at pH = 7.5. The high salinity (150 - 300 mM NaCl) in the habitat of P. modestum ensures saturation of the periplasmic Na^+ binding site and thus ATP synthesis. It is assumed that the metabolism of P. modestum is not capable of substrate level phosphorylation and hence is obligatorily dependent on the ATP provided by the ATP synthase. The ATP synthase in P. modestum does therefore not hydrolyze ATP under physiological conditions, in contrast to other bacteria which can employ the ATP synthase as primary ion pump to establish the pivotal membrane potential.

It is therefore unclear, why the affinity on the periplasmic side is much lower compared to the cytoplasmic side. It has to be assumed, that the ATP synthase was designed as enzyme capable of both functions and that the observed arrangement of different affinities of binding sites is crucial for the rotary mechanism.

In the previous chapter the transport behavior of the E. coli F_0 part was characterized and a similar situation as in P. modestum was found: Proton transport in synthesis direction displayed an apparent pK_a of 7.5 (6.2 in ATP synthesis measurements) whereas in hydrolysis direction a constant transport rate was observed between pH 6 and 9. These data are in perfect analogy with the P. modestum enzyme but the question why the affinity for the coupling ion on the periplasmic side is so low remains.

The possibility, that an additional low affinity binding site in the periplasmic access channel controls the access of Na^+ ions was declined in view of the results from the I. tartaricus cY70F mutant. This mutant is defective in Na^+ binding with an 35-fold increased K_D for Na^+ ions (von Ballmoos and Dimroth 2007). During ATP synthesis this mutant displayed an apparent K_D for Na^+ > 190 mM which favors a direct accessibility of the binding site from the periplasm. Hence, the binding sites on the c-ring seem to have a decreased affinity in the periplasmic access channel compared to the cytoplasmic access channel. We suggested that this difference is caused by the interaction of an unoccupied binding site with the positively charged arginine. To clarify how the two driving forces, ΔpNa and $\Delta \psi$, might influence this interaction which has to be broken during rotational catalysis, the impact of $\Delta \psi$ on the periplasmic Na^+ affinity in the P. modestum ATP synthase was determined.

4.3.5 $\Delta \psi$ does not affect K_D during ATP synthesis

Based on a previously published model for the F_0 part of P. modestum (Xing et al. 2004), we reasoned that $\Delta \psi$ might destabilize the interaction between aR226 and cE65, thereby facilitating binding of the Na^+ ion to the c-subunit from the periplasmic access channel. If this hypothesis was true, a varying $\Delta \psi$ would influence Na^+ binding and thus affect the K_D for Na^+ during ATP synthesis. Interestingly, the apparent K_D's remained almost constant, when ATP

4 Impact of ΔpNa on the F_0 part from Propionigenium modestum

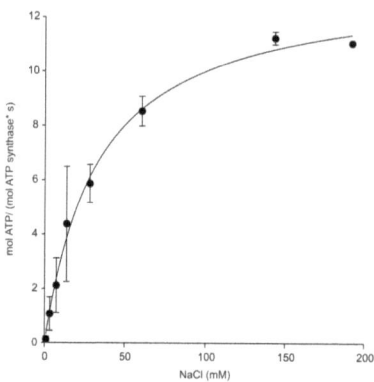

Figure 4.3: K_{app} during ATP synthesis is 35 mM. Initial synthesis rates were determined at different Na^+ concentrations and constant driving forces of $\Delta\psi$ = 150 mV and ΔpNa = 130 mV. Fitting the data with single site saturation binding yielded an apparent K_D of 35 mM.

synthesis was measured at varying $\Delta\psi$'s from 100 to 160 mV. (Panel A, Fig. 4.4 on page 77). When the Na^+ concentration inside the proteoliposomes was varied from 25 mM to 200 mM NaCl and the initial ATP synthesis rates were determined at constant ΔpNa and varying $\Delta\psi$ reduction of ATP synthesis was only observed for liposomes containing 25 mM NaCl. Under the experimental conditions binding of Na^+ apparently became rate limiting if it dropped below the apparent K_D of 35 mM during ATP synthesis.

These data imply that $\Delta\psi$ does not interfere with the aR226/ cE65 complex when located in the periplasmic access pathway. However they reinforce the idea that the Na^+ concentration is the major determinant for release of the aR226/ cE65 interaction. As a consequence, we suggest that the membrane potential exerts its main effect in the cytoplasmic channel. This view is consistent with functional studies, suggesting that the binding site is more readily accessible for hydrophilic molecules from the periplasmic than from the cytoplasmic side of the membrane (Angevine et al. 2003).

4.3.6 A model of torque generation in the *P. modestum* F_0 part in response to $\Delta\psi$ and ΔpNa

Based on experimental data presented in this and the preceding chapter a novel mechanism for torque generation by F-type ATPases is proposed with special emphasis on the different sites of action of the membrane potential and ion concentration gradient.

At the beginning of an ATP synthesis cycle, a negatively charged ion binding site on the

4.3 Results and Discussion

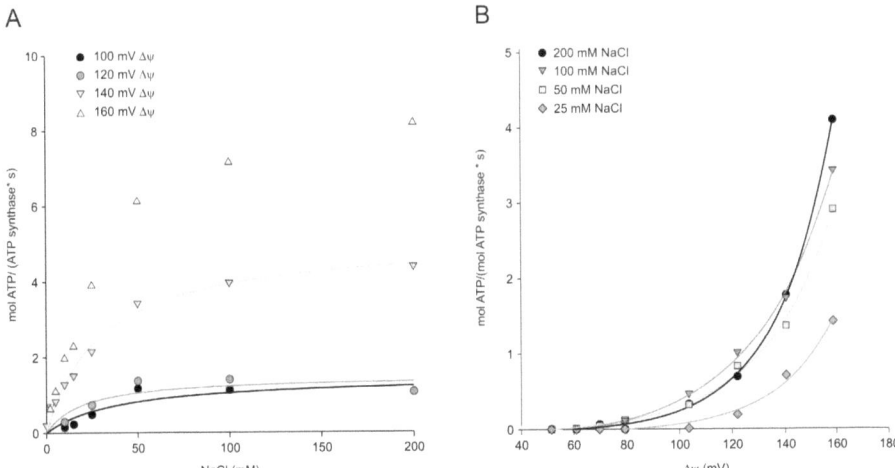

Figure 4.4: Impact of $\Delta\psi$ and Na^+ concentration on ATP synthesis. **A.** Initial rates for ATP synthesis were determined at different NaCl concentrations and with varying membrane potential ($\Delta\psi$). The apparent K_D's were unaffected by the changed $\Delta\psi$. Fitting the data yielded apparent K_D's of 39.9 mM at 100 mV, 33.8 mM at 120 mV, 30.2 mM at 140 mV and 39.2 mM at 140 mV. **B.** ATP synthesis was determined at constant ΔpNa with 4 different internal Na^+-concentrations. The $\Delta\psi$ was varied for each liposome preparation from 50 to 160 mV.

c-ring is located in the periplasmic access channel where it tightly interacts with the positively charged aR226. A coupling ion in the periplasmic channel can compete with aR226 for the binding site on the c-ring. If the coupling ion concentration is sufficiently high, the interaction is released and the occupied binding site can leave the a/ c-interface. Prerequisite for this event is the release of the Na^+ from the next incoming binding site, providing an energetically favorable charge compensation for the released aR226. Attraction between the two charges will create torque that leads to rotation of the c-ring versus the stator and ultimately to ATP synthesis in the F_1 part. The reformation of the aR226/ binding site complex creates a net rotation of the binding site towards aR226 and sets the enzyme back to the initial situation.

In this scenario, the guanidino group of aR226 is proposed to reside closer to the periplasmic than to the cytoplasmic channel in the absence of driving forces. Thus, only Na^+ ions in the periplasmic channel can directly compete with aR226 for the binding site while cytoplasmic Na^+ ions do not reach the complex. Consequently, ΔpNa can only promote rotation in synthesis direction but not hydrolysis direction, as observed experimentally. A possible structural explanation of this scenario can be deduced from recent modeling studies, which are based on a/ c cross-linking data (Vorburger et al. 2008). In the model the side chain of aR226 ap-

4 Impact of ΔpNa on the F_0 part from Propionigenium modestum

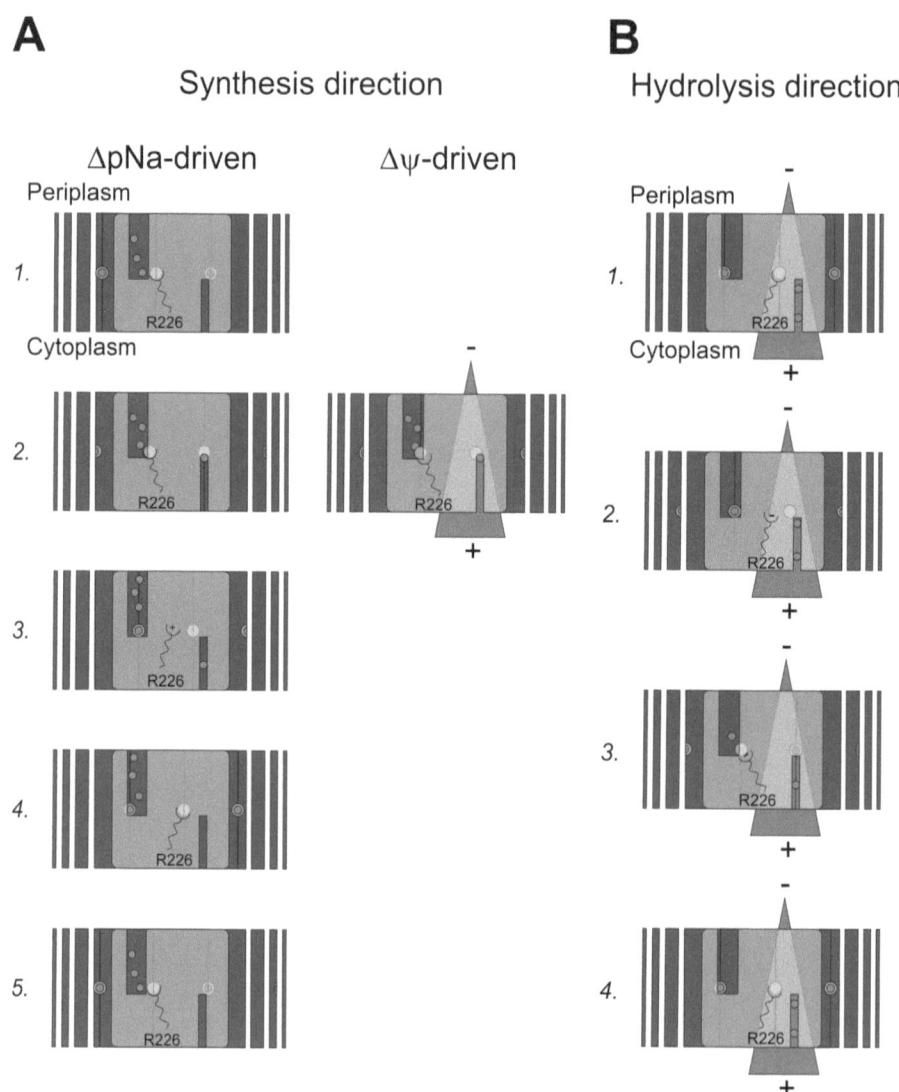

proaches the binding site from the cytoplasmic side and is therefore in a favorable arrangement for competition with the Na$^+$ ions in the periplasmic channel.

This mechanism depends critically on the Na$^+$ concentration in the periplasmic channel, which must be high enough to compete with aR226 for the binding site. In this model ΔpNa is considered to reflect the occupancy ratio of the binding sites in the periplasmic and cytoplasmic channel. During ATP synthesis, a Na$^+$ ion concentration in the periplasmic channel above the apparent K$_D$ must be accompanied by a sufficiently high probability of an empty binding site in the cytoplasmic channel to allow a net electrostatic attraction by aR226 pulling the binding site from the cytoplasmic towards the periplasmic access channel. Hence this model suggests an intimate coupling between ion binding and torque generation.

In accordance with our observations, $\Delta\psi$ is not an essential driving force for ATP synthesis in this model. It must however account for the acceleration in synthesis rate by $\Delta\psi$ which is obviously present. An effect of $\Delta\psi$ on the interaction between aR226 and the binding site in the periplasmic channel is dismissed in contrast to previous models (Xing et al. 2004) since no influence of $\Delta\psi$ on the apparent K$_D$ during ATP synthesis was observed. Consequently it is

---

Figure 4.5 *(preceding page)*: Model for torque generation in the *P. modestum* F$_0$ part. In panel A the events during **rotation in synthesis direction** are shown. **1.** aR226 is bound to cE65 (bright circle). In the absence of $\Delta\psi$ the aR226/ cE65 complex resides close to the periplasmic channel. **2.** The Na$^+$ ions (dark circles) in the periplasmic channel compete with aR226 for the binding site on the c-ring. Concomitantly the next binding site enters the cytoplasmic channel. **3.** When the Na$^+$ in the periplasmic binding site has displaced aR226 and the bound Na$^+$ ion in the cytoplasmic channel is released, the Coulomb force between the oppositely charged aR226 and cE65 will lead to rotation of the c-ring. **4.** The aR226/ cE65 complex is formed with the next c-monomer and the system is set back to the initial situation (**5.**).
In contrast to F$_1$F$_0$, the isolated F$_0$ part can rotate in synthesis direction in response to solely a $\Delta\psi$. The difference for Na$^+$-transport energized by ΔpNa and $\Delta\psi$ is described in stage **2**. In the isolated F$_0$ part the slackness of the rotor is larger due to the missing constraints imposed by F$_1$. This allows the aR226/ cE65 complex to move further into the periplasmic channel where the hydrophilic environment weakens the aR226/ cE65 interaction. Under these conditions it is possible that the aR226/ cE65 complex is released without ΔpNa. Concomitantly the Na$^+$ ion in the cytoplasmic channel is released due to $\Delta\psi$ which enables aR226 to interact with the next binding site.
In **hydrolysis direction** (panel B) transport through F$_0$ is only observed in presence of $\Delta\psi$. **1.** Due to the buried position of the aR226/ cE65 complex inside the a-subunit, the cytoplasmic Na$^+$ cannot compete for the binding site. Instead the aR226/ cE65 is disrupted by $\Delta\psi$, which requires a minimal force of 40 mV. **2.** As soon as the aR226/ cE65 interaction is released, Na$^+$ can bind to the c-subunit from the cytoplasmic channel, which is likely to be aided by $\Delta\psi$ as well. **3.** At the same time aR226 swings back towards the periplasmic channel, where an occupied binding site resides. The positive charge of the aR226 repels the bound Na$^+$ allowing the formation of a new aR226/ cE65 complex and the system can start with a new transport cycle. Similarly to ATP synthesis, torque is mainly generated by the electrostatic attraction between aR226 and the empty binding site in the periplasmic channel.

suggested that $\Delta\psi$ exerts its effects mainly in the cytoplasmic access channel by facilitating or accelerating release of bound Na^+ ions. This would promote the interaction of the binding site in the cytoplasmic channel with aR226 and in turn accelerate the synthesis rate. This idea was evoked by a cysteine accessibility study of the a-subunit which suggests a more hydrophilic periplasmic access pathway in contrast to the predominantly hydrophobic cytoplasmic channel (Angevine et al. 2003). Thus, only small potential drop occurs along the hydrophilic periplasmic channel as proposed earlier (Xing et al. 2004) but a large decrease takes place along the cytoplasmic channel, caused by its hydrophobicity. The mechanism proposed so far can explain all experimental observations during ATP synthesis for the *P. modestum* as well as the *E. coli* enzyme.

However, transport through the F_0 part responds differently to the driving forces compared to the holoenzyme. Whereas ΔpNa alone is capable of energizing ATP synthesis at low rates, $\Delta\psi$ is only effective if a ΔpNa of at least 30 mV is applied concomitantly. In contrast either ΔpNa or $\Delta\psi$ alone lead to Na^+-transport through the isolated F_0 part. This difference between F_0 and F_1F_0 is likely to be due to the constraints imposed by the F_1 part. It is therefore suggested that the isolated F_0 part shows a higher rotational flexibility which enables the R226/ binding site complex to rotate further into the periplasmic access channel. The hydrophilic environment inside the channel decreases the electrostatic interaction between aR226 which allows an interruption of the interaction in the absence of ΔpNa. In contrast to transport in synthesis direction, transport through the isolated F_0 part in hydrolysis direction requires a $\Delta\psi$ > 40 mV. No impact of ΔpNa has been observed. In the absence of driving forces we envision that the guanidinium group of aR226 in complex with the empty binding site is located close to the periplasmic channel and is not accessible from the cytoplasmic channel. This prevents cytoplasmic Na^+ ions (and thus ΔpNa) from influencing the aR226/ cE65 interaction during Na^+-transport in hydrolysis direction. In accordance with the model for ATP synthesis, it is suggested that the influence of the membrane potential is most potent in the cytoplasmic access pathway, disrupting the aR226/ cE65 complex and allowing Na^+ ions to bind from the cytoplasmic side.

While the Na^+-dependent ATP synthase from *P. modestum* and the H^+-dependent ATP synthase from *E. coli* behave very similar during ATP synthesis, differences exist in the isolated F_0 part. In the H^+-dependent ATP synthase no thresholds are observed and pH as well as $\Delta\psi$ lead to H^+-transport (see chapter 2). These variations are likely to be attributed to differences in the ion binding site.

It has recently been shown that the binding sites on the isolated c-rings from *E. coli* and *P. modestum* display fundamental differences with respect to proton binding (von Ballmoos and Dimroth 2007). Differences in the binding sites are similarly exhibited by the different effect of ethyl isopropyl amiloride (EIPA) on the ATP synthases from *E. coli* and *P. modestum*, respectively, where EIPA is supposed to mimic the guanidinium group of the stator arginine. Surprisingly only the Na^+-dependent ATP synthase is inhibited in a Na^+-dependent fashion (Vorburger et al. 2008). This implies that the interaction in the *P. modestum* enzyme between aR226 and cE65 is more specific and likely to be stronger than between aR210 and cD61 in the *E. coli* ATP synthase. Recent energy minimization calculations between a part of TMH 4 of subunit a and the c-ring favor this view (Vorburger et al. 2008). The tighter aR226/ binding site complex might thus serve as explanation why a threshold exists in the F_0 part of *P. modestum* but not in *E. coli*. Additionally, the smaller size of a proton compared to a Na^+ ion might allow a more efficient diffusion within the a/ c interface and thus a facilitated disintegration of the arginine/ binding site complex.

4 Impact of ΔpNa on the F_0 part from Propionigenium modestum

5 General Discussion - Structure and Function of the a-Subunit in F-type ATPases

The membrane embedded F_0 part of F-type ATPases is responsible for ion translocation across the membrane and torque generation during ATP synthesis. In bacterial ATP synthases the F_0 part consists of 3 different subunits with the stoichiometry ab_2c_{10-15}. Subunit b forms a homodimer (exception is a bb' heterodimer in several photosynthetic bacteria) which winds up as a coiled coil. The b_2-dimer constitutes the peripheral stalk which is believed to mainly counteract the torque generated by the rotor (Dunn et al. 2000). No direct participation in H^+-transport has been reported. However, proton conduction in the *E. coli* F_0 part was only observed in presence of the transmembrane helix of subunit b (Greie et al. 2004), which implies at least an indirect involvement in the transport process.

The c-subunits assemble as an oligomeric ring which acts as revolving barrel between the periplasmic and cytoplasmic access channels. The stoichiometry of the c-ring varies between different organisms and values from 10-15 have been described so far (Meier et al. 2007). Each c-monomer has a strictly conserved carboxylic acid buried within the membrane which is crucial for binding of the coupling ion (von Ballmoos et al. 2002a). As inferred from the X-ray structure, the c-ring seems to be rather rigid (Meier et al. 2005), showing an overall hourglass shape and a very compact packing of the monomers. The c-ring structure makes models employing large movements or helix twisting for torque generation obsolete (Aksimentiev et al. 2004; Rastogi and Girvin 1999b). Similarly, no obvious Na^+ channel from the binding site towards the cytoplasm was observed in the structure as suggested previously (Dimroth et al. 2000).

The a-subunit is the least understood but functionally most important subunit of the F_0 part. It is highly hydrophobic and consists of 5 transmembrane helices (TMHs) which seem to accomplish critical tasks required for a functional F_0 part: generation of torque together with the c-subunit in response to an electrochemical gradient and provision of access to the membrane-embedded c-ring binding sites from both the periplasm and the cytoplasm. Most importantly, subunit a forms a delicate interface with the c-ring which has to provide enough

5 General Discussion - Structure and Function of the a-Subunit in F-type ATPases

stability to guarantee integrity of the enzyme while at the same time allowing smooth rotation of the c-ring, necessary for high turnover rates and the almost 100 % efficiency of the enzyme (Yasuda et al. 1998). The a-subunit is thus perfectly tailored to huddle against the c-ring and seems to collapse if backing through the c-ring is lost e.g. by overexpression or purification of isolated subunit a. This has severely hampered *in vitro* investigations, and knowledge is very scarce, even the topology is still under debate. As a consequence, most functional investigations concerning the a-subunit were preformed in living cells or in membranes, resulting in data of rather qualitative nature which are in some cases contradictory.

This chapter aims to give an overview on the current knowledge on the a-subunit, integrating results from various studies with different approaches.

5.1 Topology of the a-subunit

In first attempts to elucidate the topology of the a-subunit, experiments were preformed using genetic fusions with alkaline phosphatase. Two independent reports predicted 8 transmembrane helices (TMHs) (Bjørbaek et al. 1990; Lewis et al. 1990). Even though both reports agree in the total number of TMHs, there is considerable variation in position and length. Moreover, the location of the termini differs between the reports.

A few years later, Futai and colleagues presented a model, in which subunit a consisted of 6 transmembrane helices and both termini resided in the cytoplasm (Yamada et al. 1996). The model was based on the reactivity of antibodies directed against predicted loop regions from hydrophobicity plots.

Alternatively, the groups of Fillingame and Vik used N-alkylmaleimide labeling of cysteine residues to determine the membrane topology of the a-subunit. Both reports predicted 5 transmembrane helices with the N-terminus in the periplasm and the C-terminus being located in the cytoplasm (Long et al. 1998; Valiyaveetil and Fillingame 1998).

This model, however, was challenged by Altendorf and colleagues who used monoclonal antibodies to map the accessibility of loop regions from either side of the membrane. Their data conflicted only in the N-terminal region, while the data agreed on the C-terminal helices. They found that two epitopes (E_4-D_{10} and V_{29}-Q_{32}) close to the N-terminus were accessible for antibodies in inside-out vesicles, concluding that the N-terminus resides in the cytosol, and the a-subunit has an even number of transmembrane helices (Jäger et al. 1998).

In 1999, Vik and colleagues confirmed their earlier data, using an antibiotic to permeabilize the outer membrane which allowed labeling of intact cells with N-alkylmaleimide (Wada et al. 1999). In an unpublished study, Yoshida and colleagues fused the C-terminus of the c-monomer to the N-terminus of subunit a of the *Bacillus* PS3 ATP synthase, inserting a Factor X_a cleavage

5.1 Topology of the a-subunit

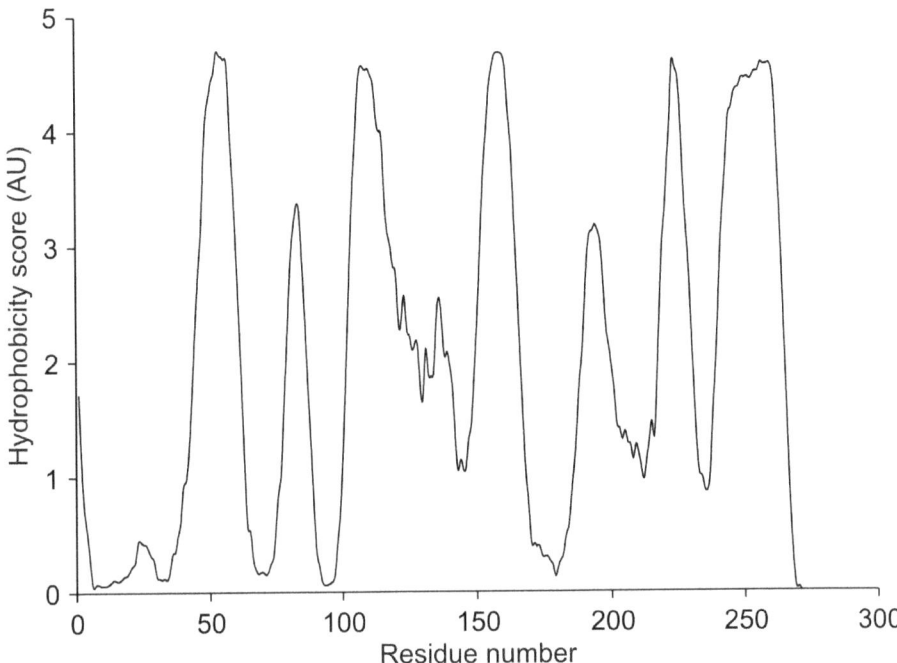

Figure 5.1: Hydrophobicity plot of subunit a from *E. coli*. Based on the amino acid sequence of the *E. coli* a-subunit a hydrophobicity plot was calculated using the SPLIT (Membrane Protein Secondary Structure Prediction Server) server. Seven TMHs are predicted, TMH 1 from 40 to 63, TMH 2 from 100 to 121, TMH 3 from 125 to 142, TMH 4 from 145 to 168, TMH 5 from 186 to 206, TMH 6 from 215 to 233 and TMH 7 from 239 to 264. Except for the TMH 4 and TMH 5 the predictions match well with experimental evidence.

site (Noryio Mitome, personal communication). The construct yielded a correctly folded ATP synthase, which is however inactive due to the covalent connection between subunits a and c. After proteolysis with Factor X_a the connection is broken and the enzyme became fully active, indicating a proper assembly of the complex. These observations confirm the suggested periplasmic residence of the N-terminus of subunit a.

Combining all models a fairly clear picture of the topology of subunit a can be drawn although some uncertainty regarding the exact beginning and end of TMHs remains (Figure 5.2). TMHs 1-3 stretch approximately from residue 38 to 62, 100 to 127, and 138 to 166, respectively. Transmembrane helices 4 and 5, covering residue 202 to 230 and 238 to 263, respectively, are similar in all models. As a consequence, the a-subunit has two long cytoplasmic loops of roughly

5 General Discussion - Structure and Function of the a-Subunit in F-type ATPases

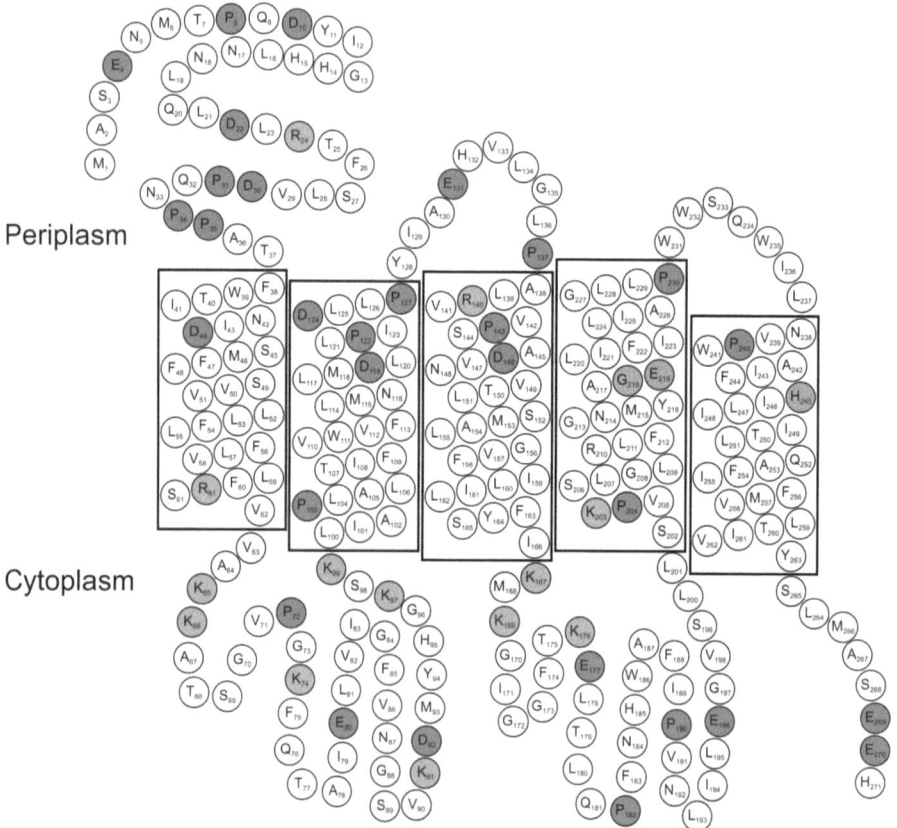

Figure 5.2: Topology model for the *E. coli* a-subunit. Acidic, basic residues, prolines and the essential R210 are labeled in different grey tones. The triad G218, E219 and H245, which is crucial for ion conduction, is marked as well.

5.2 Structural information provided by Cys-Cys cross-linking

37 amino acids in length and 2 short periplasmic loops of approximately 10 residues. The N-terminus is located in the periplasm and a short C-terminal tail in the cytoplasm. Truncation of the C-terminus after S256 did not affect function of the enzyme (Eya et al. 1991).

All models described above assume, that the a-subunit is a bundle of straight α-helices. However, subunit a contains 15 proline residues, some of them falling into the predicted transmembrane regions. It is therefore not unlikely that some of the transmembrane helices have a break within the membrane. Additionally, the length of the predicted helices ranges from 25 to 29 residues, resulting in helices with a length of 37 - 44 Å, respectively. This is unusually long for simple, straight TMHs and further supports the idea, that the structure of the TMHs in the a-subunit is more complex.

5.2 Structural information provided by Cys-Cys cross-linking

In order to obtain information about the relative assembly of the helices in an α-helical bundle, Fillingame and co-workers constructed 92 cysteine double mutants and tested them for interhelical Cys-Cys cross-link formation. Interestingly, only 7 cross-links were observed by oxidation with Cu^{2+} and one cross-link was formed after addition of iodine, all residing in a

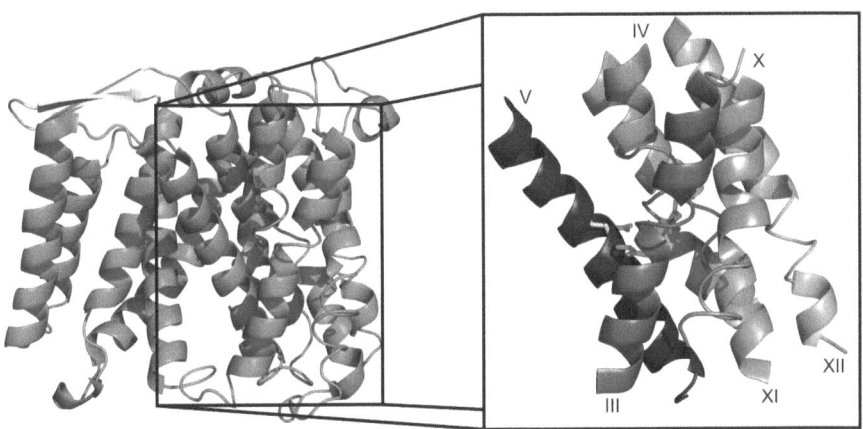

Figure 5.3: Structure of the Na^+/ H^+ exchanger NhaA from *E. coli*. On the left side the NhaA structure is shown (Hunte et al. 2005). Within the transmembrane part several unordered loops are present. On the right side the transmembrane helices of domain A consisting of TMHs III, IV, V, X, XI and XII are shown. Three aspartate residues probably involved in Na^+ binding are shown in stick representation.

5 General Discussion - Structure and Function of the a-Subunit in F-type ATPases

narrow stretch slightly above the middle of the membrane as shown in figure 5.4 (Schwem and Fillingame 2006). A possible reason for the low cross-link yield might be the limited access of Cu^{2+} ions or I_2 to the protein interior and therefore inefficient cross-link formation (hydrophilic Cu^{2+}, bulky iodine). The resulting pattern shows that only 3 residues form cross-links with a very small number of residues which are distributed on TMHs 2-5 and according to the topology model, are confined to a very narrow strip parallel to the membrane plane (Figure 5.4).

The maximal distance between two α-C atoms must be < 8 Å for a cysteine-cysteine cross-link to occur. The observed cross-links, however, would require much larger distances than 8 Å in the proposed α-helical bundle (Dmitriev et al. 2008) and make it unlikely that the a-subunit consists of 5 straight TMHs as already noted earlier.

Accordingly, the cross-linking pattern can only be explained by a certain flexibility of the residues in position 120, 218 and 248. An example for highly flexible loops within a transmem-

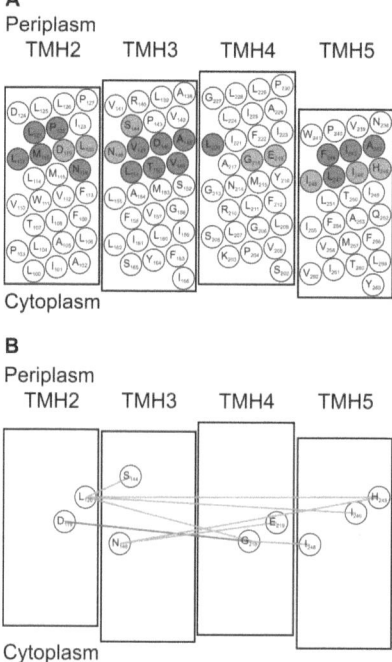

Figure 5.4: Interhelix crosslinking in the *E. coli* a-subunit. A. All residues involved in the cross-linking study by Schwem and Fillingame (2006) are highlighted. Not all combinations were tested, see Schwem and Fillingame (2006) for details. **B.** All observed cross-links are shown. The position of the residues in the membrane was adjusted to allow better presentation of the cross-links.

brane helix is observed in the structure of the Na^+/H^+ exchanger NhaA from *E. coli* (Hunte et al. 2005). The loops in the middle of TMH IV and TMH XI separate the periplasmic from the cytoplasmic funnel. It was proposed that the loops undergo fast conformational changes which enable the exceptionally fast rate of up to 1,500 s^{-1} for the Na^+/H^+ exchange between the two reservoirs. The structure of domain A of NhaA in figure 5.3 shows a 6 helix bundle located in the membrane. Most helices are tilted relative to the membrane normal and relative to each other. Flexible loops are found interspersed in TMHs IV and XI similar to our suggestion for the structure for the a-subunit deduced from the cross-linking data.

The absence of a stable secondary structure in that part of subunit a is further supported if interactions with the c-subunit are taken into account. Cross-linking experiments between subunit a and c are observed for 3 residues on TMH 2 of subunit c with 4 residues in TMH 4 of subunit a (aI221C-cG69C, aI223C-cL72C, aL224C-cY73C, aI225C-cY73C) (Jiang and Fillingame 1998). The c-subunit is known to form an α-helical hairpin and residues 69, 72 and 73 map to one side of the helix, it is therefore readily conceivable that these residues are part of the interface with the a-subunit. In contrast the 3 consecutive residues in subunit a cross-linking to subunit c make an α-helical conformation highly unlikely. In addition, the c-ring of *I. tartaricus* was shown to have a concave surface with the binding site being located in the apex. It is reasonable to assume that the a-subunit fits into this groove provided by the c-ring to allow efficient ion transfer between the subunits. To accomplish this it is very likely that at least one TMH of the a-subunit has a similar kink as observed for the *I. tartaricus* and *Bacillus* PS3 c-subunit (Nakano et al. 2006).

It is notable, that the observed stretch of cross-linkable residues contains several acidic residues (D119, D148, E219) despite their intramembranous localization. This might indicate a hydrophilic cavity in this region, supporting an intimate contact of these residues.

5.3 Cysteine scanning mutagenesis of the a-subunit

In an impressive *Tour de Force*, Fillingame and co-workers mutated every single residue of the a-subunit to cysteine, starting from the assumed beginning of TMH 1 at W39 until the C-terminus. All mutants were tested for viability (growth on glucose), ability to synthesize ATP (growth on succinate), proton pumping activity (ACMA quench) and water accessibility of the mutated cysteine (inhibition of ACMA quench after labeling with N-ethylmaleimide (NEM) or Ag^+) (Angevine and Fillingame 2003; Angevine et al. 2003 2007; Moore et al. 2008; Steed and Fillingame 2008). To probe the proton pathway the authors incubated membranes with $AgNO_3$ and NEM before estimating H^+-transport rates with ACMA quench. The ionic radius of Ag^+ ions (1.26 Å) is just in between the radius of Na^+ (0.97 Å) and H_3O^+ (1.54 Å)

5 General Discussion - Structure and Function of the a-Subunit in F-type ATPases

ions and was hence considered a suitable probe for the ion pathway. NEM in contrast is much larger and thought to indicate cavities within the transmembrane part of subunit a. However, NEM is hydrophobic and could access the cysteines from the lipid phase. The impact of NEM or Ag^+ on H^+-transport was monitored by calculation of the ratio of ACMA quenching with and without the treatment. While black and white results are readily accessible by these methods, its quantification should not be overestimated. Further, it should be noted that NEM and Ag^+ preferably react with the ionized form of the sulfhydryl group and thus no labeling is observed if the mutated cysteine residue is shielded from the external pH or the pK_a is elevated. Nevertheless, from the wealth of accumulated data by these experiments, a selection is summarized below.

In TMH 1 only D44C showed strongly impaired ACMA quench after labeling with NEM as well as with Ag^+. All other mutants of TMH 1 were fully active pointing to a predominantly structural role of TMH 1.

In TMH 2 only residue D119C, which is located approximately in the periplasmic third of the helix, stands out. The mutant displayed slightly decreased growth on succinate, severely reduced ACMA quench, strong inhibition by Ag^+ and modest inhibition by NEM. According to these results, D119 contributes to efficient ATP synthesis and proton transport but is not essential. It is tempting to speculate that the negative charge is involved in proton conduction. The periplasmic part of TMH 2 is strongly inhibited after labeling with Ag^+ which continues in the periplasmic half of TMH 3. This pattern suggests a water accessible pathway adjacent to the two helices.

Except for D146C the mutations in TMH 3 have no impact on growth, neither on glucose nor succinate. D146C is strongly inhibited after NEM labeling but only modestly inhibited after Ag^+ labeling which implies that it lines the ion pathway, but the carboxyl group does not directly contribute to proton conduction. Surprisingly, several Ag^+-sensitive cysteine mutants were found in the first two cytoplasmic loops connecting TMH 1 and TMH 2 (63-99) and TMH 3 and TMH 4 (167-201). Notably, cysteine mutants W186C - L195C showed continuously increasing inhibition by Ag^+. This pattern is different from most other Ag^+ sensitive stretches and difficult to reconcile with an α-helical structure. On the other hand the Ag^+ inhibition in the second cytoplasmic loop can only be explained by assigning a function in H^+ translocation. NEM labeling is neither observed for the first nor the second cytoplasmic loop. Interestingly both cytoplasmic loops could be cross-linked with methanethiosufonate (MTS) derivatives albeit no Cys-Cys cross-links could be observed (Moore, Angevine et al. 2008). The two loops could possibly interact and form a single domain which harbors the entrance for the cytoplasmic access channel.

In TMH 4, residues 206, 210 and 214 display a high degree of NEM sensitivity as well as

5.3 Cysteine scanning mutagenesis of the a-subunit

pronounced inhibition of ACMA quenching by Ag^+, implying that R210 is surrounded by a hydrophilic environment. The mutant E219C is not inhibited with NEM (Angevine et al. 2003) but is strongly impaired in growth on succinate. The effect of Ag^+ labeling is difficult to interpret as ACMA quench is already drastically reduced in the unlabeled mutant. It is however evident that the negative charge of E219 is crucial for efficient proton translocation.

In TMH 5 only Q252C is impaired in ACMA quenching while growth on succinate is comparable to the wildtype. After Ag^+ labeling, ACMA quenching of the mutants I248C and Q252C is strongly inhibited. Both residues are presumably in close proximity to R210 as inferred from the second site revertant R210Q/ Q252R (Hatch et al. 1995). Finally, a strong effect of NEM-labeling is observed for V262C and with Ag^+ for the neighboring Y263C mutant. Both residues lie at the cytoplasmic terminus of the TMH 5.

In general, results of ACMA quenching in membranes only poorly correlate with synthesis activity deduced from growth on succinate. This might be due to the fact that ACMA quenching is not quantitative (Dencher et al. 1986) and that differences in growth behavior on succinate between different groups exists. Moreover, it can not be excluded that proton translocation in synthesis (during growth on succinate) and hydrolysis mode (as in ATP-driven ACMA quench) are not equivalent as observed in several mutants described by Howitt et al. (1990).

In summary the experiment of Fillingame's group identified a limited set of residues whose mutation to cysteine severely affected either growth on succinate or proton conduction. These residues are D119 in TMH 2, V141 in TMH 3, Q252 in TMH 5 and L207, G208, R210 G213, E219 in TMH 4. Only few residues displayed a strong and clear impact on ACMA quenching after labeling with NEM: D44C, D146C, S206C, N214C and V262C. Whereas S206 and V262 are located at the cytoplasmic end of TMH 4 and 5, respectively, D44 and D146 are in the periplasmic half of TMH 1 and 3, respectively. As these residues show only impaired proton conduction after NEM labeling, the charges of D44 and D146 seem not to be functionally important by themselves. However, when both D44 and D124 are mutated concomitantly enzyme activity is lost. Together with D119, E218 and H245, these aspartates could form a cluster of charged and hydrophilic amino acids on the periplasmic surface of subunit a which might facilitate proton capture and/ or conduction towards the binding site on the c-ring. The pattern of Ag^+ sensitive residues is fairly complex. The periplasmic halves of TMH 2 and TMH 3 exhibit a high degree of labeling and inspired by this observation Fillingame and co-workers mapped the periplasmic access channel to this interface. Another stretch of highly Ag^+ sensitive residues is found at the cytoplasmic end of TMH 4 and at residues around Y263. Surprisingly, Ag^+ labeling in the first and second cytoplasmic loops showed impact on H^+-conduction. It was therefore proposed that the cytoplasmic loops might fold into a single domain that harbors the entrance of the cytoplasmic channel (Moore et al. 2008).

5.4 Functional mutations in the *E. coli* a-subunit

During the past 20 years, about 300 different mutations were introduced in the *E. coli* a-subunit and characterized in various laboratories (see table 5.1). In most cases growth on succinate was used as indicator if the ATP synthase is functional. Due to varying incubation times and conditions between laboratories, the results differ for several mutants. Proton transport in hydrolysis direction was investigated in inverted vesicles using ATP-driven ACMA quench whereas proton conduction in synthesis direction was assessed by measuring NADH-driven H^+ backflow in membranes devoid of F_1.

Valuable information are provided by second site suppressor mutations (see table 5.2). These mutations can rescue an otherwise inactive enzyme and thereby provide insights into spatial arrangements and functional requirements of the respective residues. In view of the vast number of mutants not all results can be discussed in detail. The mutants are grouped according to their effects on the ATP synthase: 1.) Essential residues in the ATP synthase. 2.) Mutants leading to impaired growth on succinate. 3.) Second site suppressor mutations.

Table 5.1: Effect of single mutations in the *E. coli* a-subunit.

mutation	description	reference
H15D	very poor growth on succinate	Patterson et al. (1999)
	interrupts assembly/ insertion of subunit a	
D44N	normal growth on succinate	Howitt et al. (1990)
	ACMA quench with F_1F_0 but not F_0	
R61N	no effect	Howitt et al. (1988)
K74C	cross-links with b-subunit	Long et al. (2002)
	cross-link does not affect function	
E80K	50 % specific activity of wt	Long et al. (2002)
	reduced proton permeability	
E80A	73 % specific activity of wt	Long et al. (2002)
E80L	76 % specific activity of wt	Long et al. (2002)
E80Q	wt activity	Long et al. (2002)
D119H	suppressor of H245C	Valiyaveetil and Fillingame (1998)
D119C	slow groth on succinate	DeLeon-Rangel et al. (2003)
	good ACMA quench	
D119A	almost like wt	Howitt et al. (1990)
D119H	reduced ACMA quench	Paule and Fillingame (1989)
D124N	normal growth on succinate	Howitt et al. (1990)

5.4 Functional mutations in the E. coli a-subunit

Table 5.1: continued

mutation	description	reference
R140Q	ACMA quench with F_1F_0 but not F_0 normal growth on succinate	Howitt et al. (1990)
P143S	ACMA quench with F_1F_0 but not F_0 slow growth on succinate	Eya et al. (1988)
D146N	impaired H^+-conduction almost like wt	Howitt et al. (1990)
S152F	impaired proton conduction	Paule and Fillingame (1989)
K167L	no effect	Lightowlers et al. (1987)
K169L	no effect	Lightowlers et al. (1987)
Q181H	no effect	Vik et al. (1990)
D184Y	no effect	Vik et al. (1990)
D184H	no effect	Vik et al. (1990)
H185Y	no effect	Vik et al. (1990)
H185Q	no effect	Vik et al. (1990)
P190R	no proton translocation	Vik et al. (1988)
P190Q	no proton translocation	Vik et al. (1988)
P190N	almost no effect	Vik et al. (1988)
N192L	growth on succinate	Vik et al. (1990)
N192V	growth on succinate	Vik et al. (1990)
N192P	slow growth	Vik et al. (1990)
N192S	growth on succinate	Vik et al. (1990)
N192T	growth on succinate	Vik et al. (1990)
N192R	slow growth	Vik et al. (1990)
G197R	impaired growth on glucose	Vik et al. (1988)
E196D	almost no ACMA quench no effect	Paule and Fillingame (1989) Vik et al. (1988)
E196Q	no effect	Vik et al. (1988)
E196H	no effect	Vik et al. (1988)
E196N	no effect	Vik et al. (1988)
E196K	no growth effect, 50 % ACMA quench	Vik et al. (1988)
E196A	no effect	Vik et al. (1988)
E196S	no effect	Vik et al. (1988)
E196P	no growth on succinate	Vik et al. (1988)

Table 5.1: continued

mutation	description	reference
S199T	30 % ACMA quench no effect	Vik et al. (1988)
S199A	no effect	Vik et al. (1988)
		Howitt et al. (1988)
S202A	no effect	Howitt et al. (1988)
K203I	no effect	Lightowlers et al. (1988)
S206A	no effect	Howitt et al. (1988)
S206L	slow growth on succinate reduced ACMA quench	Cain and Simoni (1986)
L207C	no effect	Cain and Simoni (1989)
L207Y	no effect	Cain and Simoni (1989)
L207R	NARP[1] mutation, inactive	Hartzog and Cain (1994)
L207P	NARP mutation, inactive	Hartzog et al. (1999)
R210Q	no proton conduction	Lightowlers et al. (1987) Eya et al. (1991)
R210K	inactive	Cain and Simoni (1989) Eya et al. (1991)
R210I	inactive	Cain and Simoni (1989)
R210V	inactive	Cain and Simoni (1989)
R210E	inactive	Cain and Simoni (1989)
L211Y	normal growth on succinate	Cain and Simoni (1989)
L211F	normal growth on succinate	Cain and Simoni (1989)
G213C	growth on succinate ($\sim$50 % of wt) hardly any ACMA quench	Kuo and Nakamoto (2000)
G213S	almost like wt	Kuo and Nakamoto (2000)
G213T	ACMA quench & backflow 50 % of wt	Kuo and Nakamoto (2000)
G213L	similar to G213C	Kuo and Nakamoto (2000)
G213N	slow growth on succinate no ACMA quench	Kuo and Nakamoto (2000)
N214V	normal growth on succinate	Cain and Simoni (1989)
N214L	very slow growth on succinate	Cain and Simoni (1989)

---

[1]Neuropathy, ataxia, and retinitis pigmentosa; a hereditary disease caused by a mutation in the human atp6 gene.

Table 5.1: continued

mutation	description	reference
N214Q	normal growth on succinate	Cain and Simoni (1989)
N214H	no growth on succinate	Cain and Simoni (1989)
N214E	normal growth on succinate	Cain and Simoni (1989)
A217H	very slow growth on succinate	Cain and Simoni (1989)
	50 % less ACMA quench	
A217R	inactive	Cain and Simoni (1989)
A217L	very slow growth on succinate	Cain and Simoni (1989)
	50 % less ACMA quench	
G218A	normal growth on succinate	Cain and Simoni (1989)
G218V	very slow growth on succinate	Cain and Simoni (1989)
	poor ACMA quench	
G218D	very slow growth on succinate	Cain and Simoni (1989)
	good ACMA quench	Hartzog and Cain (1994)
		Gardner and Cain (1999)
G218K	no proton translocation	Hartzog and Cain (1994)
E219H	slow growth on succinate	Howitt et al. (1990)
	no ACMA quench	Cain and Simoni (1988)
		Eya et al. (1991)
E219Q	no growth on succinate	Lightowlers et al. (1988)
		Cain and Simoni (1988)
		Eya et al. (1991)
E219A	slow growth on succinate	Lightowlers et al. (1988)
		Valiyaveetil and Fillingame (1997)
E219D	normal growth on succinate	Cain and Simoni (1988)
	ACMA quench like wt	Valiyaveetil and Fillingame (1997)
E219L	no growth on succinate	Cain and Simoni (1988)
	no ACMA quench	
E219C	no growth on succinate	Valiyaveetil and Fillingame (1998)
	50 % less ACMA quench	
E219K	normal growth on succinate	Valiyaveetil and Fillingame (1997)
	50 % less ACMA quench	
E219G	reduced growth on succinate	Valiyaveetil and Fillingame (1997)
	ACMA quench like wt	

Table 5.1: continued

mutation	description	reference
P230L	50 % growth on succinate	Eya et al. (1991)
	poor ACMA quench	
W231-end	inactive	Eya et al. (1991)
		Paule and Fillingame (1989)
P240L	no effect, suppressor of bG9D	Kumamoto and Simoni (1986)
P240A	no effect, suppressor of bG9D	Kumamoto and Simoni (1986)
H245G	very slow growth on succinate	Vik et al. (1998)
		Hartzog and Cain (1994)
H245S	no growth on succinate	Vik et al. (1998)
H245C	no growth on succinate	Vik et al. (1998)
H245L	inactive	Lightowlers et al. (1987)
H245E	slow growth on succinate	Cain and Simoni (1988)
	ACMA quench & backflow 50 % reduced	Eya et al. (1991)
H245Y	no proton conduction through F_0	Cain and Simoni (1986)
Q252E	growth on succinate	Vik and Antonio (1994)
	good ACMA quench	Eya et al. (1991)
Q252H	growth on succinate	Vik and Antonio (1994)
Q252S	growth on succinate	Vik and Antonio (1994)
Q252N	growth on succinate, leaky F_0 part	Vik and Antonio (1994)
Q252C	growth on succinate	Vik and Antonio (1994)
Q252V	growth on succinate	Vik and Antonio (1994)
Q252L	growth on succinate	Vik and Antonio (1994)
	poor ACMA quench	Eya et al. (1991)
Q252G	no growth on succinate	Vik and Antonio (1994)
Q252Y	no growth on succinate	Vik and Antonio (1994)
Q252F	no growth on succinate	Vik and Antonio (1994)
Q252I	no growth on succinate	Vik and Antonio (1994)
Q252W	no growth on succinate	Vik and Antonio (1994)
Q252K	no growth on succinate	Vik and Antonio (1994)
Q252R	no growth on succinate	Vik and Antonio (1994)
Y263X	all aa's[2] except D & N	Vik et al. (1991)
	all viable on succinate	

---

[2] aa = amino acid

5.4 Functional mutations in the E. coli a-subunit

Table 5.1: continued

mutation	description	reference
	H$^+$ translocation with non-polar aa's	
	impaired H$^+$ translocation with polar aa's	
Y263F	good growth on succinate	Eya et al. (1991)
Y263-truncation	no growth on succinate	Eya et al. (1991) Vik et al. (1991)
S265-truncation	like wt	Eya et al. (1991)

5.4.1 Essential residues in the ATP synthase

The most conserved residue in the ATP synthase is arginine 210. Every mutation of the arginine led to a complete loss of activity. In current models of the ATP synthase, the positive charge is believed to ensure release of the coupling ion from the binding site on the c-ring when it passes R210 as it moves from the cytoplasmic to the periplasmic access channel (during ATP synthesis) or vice versa (during ATP hydrolysis) (Elston et al. 1998; Xing et al. 2004). In addition, R210 is likely to be a crucial factor for stabilizing the a/ c interaction by transient ionic interactions between aR210 and cD61.

A further critical residue is glutamate 219. If E219 is changed to L or C no growth on succinate is observed any more. Mutations to H, Q, A or G lead to drastically reduced growth rates. Only mutations to K or D displayed growth rates on succinate comparable to wildtype. It is difficult to explain why a mutation to H with a pK$_a$ in the neutral range is not tolerated whereas the E219K and E219D mutants with either acidic or basic pK$_a$s behave almost like the wildtype enzyme.

The described mutations of H245 either led to a complete loss of growth on succinate (H245S, C, L, Y) or to reduced growth (H245G, E). However, function of H245G and H245C could be restored by replacing G218 by K or D and D119 to H, respectively. Apparently a protonatable residue close to this position is essential for enzyme function.

In addition to residues E219 and H245, G218 is supposed to be part of a triad which is critical for ion conduction. The residues are not strictly conserved between different species but rather show specific adaptations to the environment of the ATP synthase (McMillan et al. 2007; von Ballmoos et al. 2008).

5 General Discussion - Structure and Function of the a-Subunit in F-type ATPases

5.4.2 Mutants leading to impaired growth on succinate

A fairly large number of mutations led to impaired growth on succinate and resulted in decreased proton conduction. Whereas for some residues a functional participation in proton conduction can be inferred, others almost certainly have an unspecific, structural effect. This applies especially for mutations involving proline residues. Mutation of P143, P190 and P230 led to complete inactivation of the enzyme or severely impaired growth on succinate even though P190 and P230 are located in loop regions. Mutating P190 which resides in the second cytoplasmic loop, to R or Q completely abolishes growth on succinate. These are further indications that the cytoplasmic loops are well structured and functionally important.

For efficient proton capture and conduction, charged residues are crucial. Interestingly, none of the basic residues investigated (K74, K167, K169, K203, R61, R140) except for R210 was essential for enzyme function. From the acidic residues that were characterized (D44, D92, D119, D124, D146, D184, E80, E196 and E219) only D119, E196 and E219 were sensitive to mutations. Residue E196 was insensitive to a large number of mutations (D, Q, H, N, K, A, S). If a proline was placed at position 219, no growth was observed and proton conduction was strongly impaired, which is likely to be due to a disrupted structure. Decreased growth was similarly observed if D119 was mutated to C or H.

The aspartate residues D44, D119, D134 and D146 are located between the periplasmic termini of the TMHs and the triad G218/ E219/ H245 according to the topology model presented in figure 5.2. These aspartates are thus attractive candidates for creating a proton pathway towards the binding site on the c-ring. However, single mutations of any of the aspartates did not affect growth on succinate. Interestingly the double mutation D44N/ D124N rendered the F_0 part proton impermeable (Howitt et al. 1990). Similarly the cysteine mutants D44C, D119C and D146C showed only an effect if they were additionally modified by NEM or Ag^+. Taken together, the data support the idea that these aspartate residues (D44, D124, D119 and D146) form a redundant proton pathway towards binding site on the c-ring. As the D44N/ D146N double mutant is unable to pump protons during ATP hydrolysis a role exclusively in capturing protons is unlikely and a function in proton conduction is suggested. A number of uncharged but hydrophilic residues show an impact on proton conduction as well. The effects can either be caused by different space requirements of the mutated residue or by disruption of hydrogen bonds. Whereas S152F, S206L and N192P/ R most likely fall into the first category, the N214H mutation, which lies one helix turn apart from R210, might disturb the hydrogen bonding or interfere with proton shuttling. Surprisingly, most mutations of the small, aliphatic residues, which have an impact on function of the ATP synthase, locate to a stretch of 20 amino acids at the transition from the second cytoplasmic loop to the fourth TMH. Mutants with impaired

5.4 Functional mutations in the E. coli a-subunit

growth on succinate include G197R, L207R/ P, G213N/ C/ T/ L, A217H/ R/ L and G218V/ D/ K. It is very likely that the effects of these mutations are caused through steric blockage of the proton pathway or in the case of L207P disruption of the helical hydrogen-bonding pattern.

5.4.3 Second site suppressor mutations

Valuable insights were obtained from second site suppressor mutations which allow conclusions about the spatial relationship and the functional requirements of residues in a specific location. The double mutant R210Q/ Q252R is the only known replacement of R210 which is able to grow on succinate, albeit at a much slower rate (Hatch et al. 1995). The functional complementation of the otherwise inactive R210Q mutant implies that both residues are located in close vicinity.

Several other important second site suppressor mutations involve the already mentioned triad E219, H245 and G218. Mutations of H245 severely affect enzyme function. Activity of the H245C mutant could be restored by concomitantly mutating D119H. This functional compensation implies a close proximity of residues 119 and 245 which is confirmed by recent cross-linking data (Schwem and Fillingame 2006). Moreover it suggests the requirement for histidine at this position. Surprisingly, activity of the H245G mutant is restored by concomitantly mutating G218 to either K or D. Neither the carboxyl group of E219 nor of D119 is sufficient to restore growth. The reason for choosing a D and K replacement instead of G is that the H245G/ G218K mutant mimics the situation in alkaliphilic bacteria whereas the H245G/ G218D mutant mimics the situation in chloroplasts. Mutant E219C regained the ability to grow on succinate if combined with the mutation A145E. It is very likely that E145 is a functional replacement for E219. Therefore the N-terminal parts of TMH 3 and TMH 4 must be in close proximity. Interestingly, activity of the mutant E219H is restored by mutating R140 to either L or H. The impact of these second site mutations is likely to be caused by changes in the spatial arrangement rather than functional compensation. If a helical secondary structure is assumed, R140 lies almost on the opposite side of the helix and approximately 7.5 Å apart from residue 145 mentioned above. Additionally, it is not easily conceivable that a leucine residue can functionally compensate an acidic side chain as in the E219H/ R140L mutant.

Combining the results from the second site suppressor mutations a close interaction of TMH 2 (D119), TMH 3 (A145), TMH 4 (G218, E219) and TMH 5 (H245) can be presumed. This result is congruent with the cross-linking data from Schwem and Fillingame (2006) and reinforces a close interaction of TMHs 2-5 in the periplasmic part of the enzyme. The data further imply an essential role of the triad G218/ E219/ H245 in proton delivery to the c-ring. If the distances in TMH 4 and TMH 5 are approximated on the basis of the topology predictions, R210 would be located directly opposite Q252, both being 11 or 12 amino acids apart from the suggested

5 General Discussion - Structure and Function of the a-Subunit in F-type ATPases

beginning of the membrane border. Residue H245 would be situated two helix turns towards the periplasm with respect to Q252 and R210. This makes H245 a very interesting candidate for delivering protons directly to cD61. As the pK_a of histidine is higher than the pK_a of aspartate this would explain how the protons are transferred between subunit a and c. However, this hypothesis is highly speculative because i) the exact topology is unknown, ii) it seems unlikely that all TMHs are straight helices, iii) the pK_a's inside a protein can deviate significantly from the pK_a's observed in solution and iv) alkaliphilic bacteria have a lysine instead of a histidine which makes the proton transfer difficult to explain on basis of the pK_a difference.

5.4 Functional mutations in the E. coli a-subunit

Table 5.2: Described double mutants in the E. coli a-subunit.

1st mutation	2nd mutation	effect	reference
D44N	D124N	inactive	Howitt et al. (1990)
	R140Q	inactive	Howitt et al. (1990)
K65Q	K66Q	like wildtype	Howitt et al. (1990)
K97Q	K99Q	like wildtype	Howitt et al. (1990)
D119H	H245C	rescues H245 mutant	Valiyaveetil and Fillingame (1998)
D124N	R140Q	like wildtype	Howitt et al. (1990)
R210K	Q252R	complementation	Hatch et al. (1995)
R210A	N214R	no growth on succinate, H^+-conduction in F_0	Langemeyer and Engelbrecht (2007)
G213N	L251V	better growth compared to G213N	Kuo and Nakamoto (2000)
G218D	H245G	complementation - chloroplast sequence	Hartzog and Cain (1994)
G218K	H245G	complementation - alkaliphile sequence	Hartzog and Cain (1994)
E219H	R140H	normal growth on succinate	Howitt et al. (1990)
	R140L	normal growth on succinate	Howitt et al. (1990)
	R140H	no ACMA quench with F_0	Cain and Simoni (1988)
E219N	R140H	inactive	Howitt et al. (1990)
E219C	A145E	complementation	Valiyaveetil and Fillingame (1998)
Q252E	E219D	good growth on succinate	Vik and Antonio (1994)
	E219K	good growth on succinate	Vik and Antonio (1994)
	E219G	good growth on succinate	Vik and Antonio (1994)
	E219A	good growth on succinate at RT	Vik and Antonio (1994)
	E219S	good growth on succinate at RT	Vik and Antonio (1994)
	E219P	very slow growth at RT	Vik and Antonio (1994)
	E219V	very slow growth at RT	Vik and Antonio (1994)
	E219L	no growth on succinate	Vik and Antonio (1994)
	E219I	no growth on succinate	Vik and Antonio (1994)
	E219T	no growth on succinate	Vik and Antonio (1994)
	E219C	no growth on succinate	Vik and Antonio (1994)
	E219F	no growth on succinate	Vik and Antonio (1994)
	E219R	no growth on succinate	Vik and Antonio (1994)
	E219Q	no growth on succinate	Vik and Antonio (1994)

5 General Discussion - Structure and Function of the a-Subunit in F-type ATPases

5.5 Sequence alignment of bacterial, mitochondrial and chloroplast a-subunits

To obtain more information about the importance and possible function of single residues we aligned the sequences from bacterial, mitochondrial and chloroplast a-subunits. The search for sequences of subunit a from different organisms was preformed using Prosite database[3] . The searchstring used for scanning was derived from the sequence immediately C -terminal of R210 which is highly conserved. It should be kept in mind that with the used search string[4] a strict conservation of R210, L211, N214 and Q252 is presumed.

A first striking observation was that sequences from chloroplast a-subunits are almost identical. Mitochondrial sequences as well are more conserved than bacterial ones, but by far not as high as in chloroplasts. The reason for this high conservation in chloroplasts is unknown. The largest sequence variations were observed near the N -terminus whereas the most conserved sequences are found in TMH 4 and TMH 5 which harbor many functionally important residues. Interestingly, the residues found to be crucial in mutagenesis experiments, G218, E219 and H245, display a very interesting pattern, depending on the environment of the ATP synthase.

5.5.1 Sequence variations of the triad 218/ 219/ 245

Specific sequence adaptations were found in the ATP synthases of alkaliphilic bacteria (Ivey and Krulwich 1992; Wang et al. 2004). One of these adaptations is a lysine in position 218 (*E. coli* numbering) which is exclusively found in alkaliphilic bacteria. The triad described before is complemented by a glutamate in position 219 and glycine at position 245. In contrast, all analyzed a-subunits from chloroplast and cyanobacterial ATP synthases contained D218, E219 and G245, whereas in the mitochondrial sequences the triad consisted of G218, H219 and L/ Q245. There is considerably more sequence variation in other bacteria: Sequences from the genus *Bacillus* have the triad G218, E219 and S245 in common. In most other bacteria the triad consists of residues G218, H219 and E245, whereas the positions of H and E can be exchanged. In Na^+-dependent ATP synthases only uncharged residues are found in the triad: G218 M/ L/ F/ G219 and A/ V/ L245. In view of the proposal that the triad might form a transfer gate for coupling ions from the a-subunit to the c-ring, the absence of charged residues in Na^+-dependent ATP synthases might prevent the binding of protons to the conserved carboxylic acid on the c-ring.

Alkaliphilic bacteria have the problem of capturing protons at the high pH values (up to

---

[3]http://www.expasy.ch/prosite/
[4]R-L-[FYAT]-[AG]-N-X(20,60)-[ILV]-Q-[ASGT]-[LYF]-[IV]-F

Table 5.3: **Sequence variations in the triad 218/ 219/ 245.** (*E. coli* numbering.)

Source	218	219	245
chloroplasts & cyanobacteria	D	E	G
mitochondria	G	H	L/ Q
most bacteria	G	H	E
	G	E	H
genus *Bacillus*	G	E	S
Na^+	G	M/ F/ L/ G	A/ V/ L

pH 10.5). A lysine with its pK_a of approx. 10.5 would be ideally suited to recruit protons from the bulk and deliver them to the ATP synthase. Similarly in chloroplasts the thylakoid lumen has an pH < 6 under illumination (Renganathan et al. 1993) which would make an acidic residue an attractive candidate for capturing protons. Finally, in mitochondria or neutrophilic bacteria, a histidine residue would be well suited to accomplish this task. A hypothesis how the lysine might assist in capturing protons in alkaliphilic bacteria is given in the next paragraph. The sequence of *Helicobacter pylori*, which thrives in the stomach at a pH of ∼ 3 (Matin et al. 1996), has the triad sequence D218, D219, L245, reinforcing the hypothesis that the pK_a of the residues in the sequence of the triad is a major determinant for pH adaptation.

5.6 Concept of lateral proton diffusion

Alkaliphilic bacteria have to cope with an inverse pH gradient which causes the steady state proton-motive force in some cases to be smaller than the phosphorylation potential (ΔG_p). Still they can thrive on non-fermentable carbon sources (Krulwich and Guffanti 1992). An explanation how to bridge the gap between proton-motive force and ΔG_p is that proton delivery to the ATP synthase might be accomplished by lateral proton diffusion (Heberle et al. 1994). The concept of lateral proton diffusion assumes that protons which are pumped over the membrane first interact with the negatively charged phospholipid headgroups before they equilibrate with the bulk. This leads to an increased proton concentration on the membrane surface in combination with fast diffusion rates along the lipid bilayer, as shown experimentally (Heberle et al. 1994). This phenomenon could resolve the apparent discrepancies between the static proton-motive force determined for alkaliphilic bacteria and the proton-motive force required for ATP synthesis. Besides this mechanism to endure the high pH, alkaliphilic bacteria have evolved a number of specific adaptations in ATP synthases to cope with the adverse conditions (Wang et al. 2004). In the a-subunit a lysine is found in the position corresponding to G218 in *E. coli*, which was suspected to capture protons at high pH. Recently, Cook and co-workers demonstrated for the thermoalkaliphilic *Bacillus* TA2 that only with K or R in the respective position of the a-subunit synthesis was possible at high pH, while mutations to H or G lowered the pH optimum from 9.5 to 8 or 7.5, respectively (McMillan et al. 2007). Furthermore, in the H/ G mutant, synthesis was still observed at pH 6.5, which was completely absent in the wildtype or in the mutant carrying a lysine. Hence the residue at position 180 can sense the external pH and McMillan et al. (2007) suggested a water-exposed position for the lysine enabling proton capture from the bulk. However, according to membrane topology predictions for the *E. coli* a-subunit residue 218 is located in the middle of TMH 4 which makes direct water accessibility rather unlikely. Another explanation for the requirement of lysine could be that it is fairly independent from the external pH and acts as kinetic accelerator for proton conduction. Similar to D96 in bacteriorhodopsin it would ensure efficient loading of the binding site on the c-ring at high pH values (Cao et al. 1993), whereas the proton delivery would be accomplished by lateral proton transfer along the membrane surface. It is however unclear how the proton is transferred from the basic lysine to the acidic binding site on the c-ring. A possible mechanism to solve this conundrum involves the transition of the lysine from a hydrophilic environment with high pK_a into a hydrophobic pocket inside the protein which is accompanied by a decrease of the pK_a which would allow release of the proton towards the binding site on the c-ring (Ch. von Ballmoos, P. Dimroth, personal communication). The energy for the transition from a hydrophobic to a hydrophilic environment would be provided by the membrane potential which exerts its effect only on the charged (i.e. protonated) form of the lysine. This mechanism is reminiscent of the role suggested for E242 in the bovine cytochrome c oxidase (Wikström and Verkhovsky 2007). In the crystal structure, E292 is located on the bottom of the D pathway approximately in the middle of the membrane (Tsukihara et al. 1996). To accomplish its function as H^+ donor a rotation relative to the position in the crystal structure was suggested. Associated with the rotation is a drastic change of the pK_a by 2.5 pK units which was experimentally confirmed by FTIR spectroscopy (Gorbikova et al. 2007).

5.7 Outlook

The proton pathway inside the a-subunit is still elusive. Site-directed mutagenesis has revealed important residues for the proper function of the ATP synthase, however no efforts have been made to clarify the mechanistic role of any of these residues. Further mutagenesis studies which make use of data that is already available will be required to shed light into the proton pathway in the ATP synthase which is an indispensable prerequisite for elucidation of the mechanism of the F_0 rotor and the ATP synthase. From the 271 residues of subunit a in *E. coli*, only R210 is ever found in any model. This underlines how indispensable further characterization of specific residues is required in order to get a reasonable model for proton conduction and torque generation in F-type ATPases. Our findings in chapter 2 and 3 make it further obvious, that ion transport in opposing directions might be different and must therefore be clearly defined during experiments.

5 General Discussion - Structure and Function of the a-Subunit in F-type ATPases

Appendix I

Derivation of the amount of external and internal pyranine

For the following considerations, fluorescence values denoted F_{tot}, F_i and F_{ex} correspond to measured, internal and external fluorescence with the associated fluorescence ratios R_{tot}, R_i and R_{ex}, respectively. The further indices indicate the excitation wavelength and the pH of the assay buffer

$$F_{tot} = F_i + F_{ex} \tag{5.1}$$

To determine the extent of external pyranine, we measured the fluorescence of a liposome preparation (pH 7.2) in buffer with pH 7.2 and in buffer with pH 6.8. In the latter case, only the external pyranine will rapidly respond to the pH change. This allowed determination of the fraction x_{in} of internal pyranine in a liposome preparation. When measured in buffer with pH 7.2, ΔpH = 0 and the following equations are valid:

$$F_{tot(405, pH\,7.2)} = F_{ex(405, pH\,7.2)} + F_{i(405, pH\,7.2)} \tag{5.2}$$

$$F_{tot(460, pH\,7.2)} = F_{ex(460, pH\,7.2)} + F_{i(460, pH\,7.2)} \tag{5.3}$$

$$R_{tot} = R_{ex} = R_i = \frac{F_{tot(405, pH\,7.2)}}{F_{tot(460, pH\,7.2)}} = \frac{F_{a(405, pH\,7.2)}}{F_{a(460, pH\,7.2)}} = \frac{F_{i(405, pH\,7.2)}}{F_{i(460, pH\,7.2)}} \tag{5.4}$$

When the external pH is shifted to 6.8, the internal pH will remain unchanged and only the fluorescence of the external pyranine will change.

$$F_{tot(405, pH\,6.8)} = F_{ex(405, pH\,6.8)} + F_{i(405, pH\,7.2)} \tag{5.5}$$

$$F_{tot(460, pH\,6.8)} = F_{ex(460, pH\,6.8)} + F_{i(460, pH\,7.2)} \tag{5.6}$$

The fluorescence from the internal pyranine can be calculated as a fraction x_{in} of the measured fluorescence F_{tot} at a given pH value:

$$F_{i(405, pH\,7.2)} = F_{tot(405, pH\,7.2)} \times x_{in} \tag{5.7}$$

$$F_{i(460, pH\,7.2)} = F_{tot(460, pH\,7.2)} \times x_{in} \tag{5.8}$$

The internal fluorescence $F_{i(pH\,7.2)}$ in (5) and (6) can be replaced by the terms given in (7) and (8), respectively:

$$F_{tot(405, pH\,6.8)} = F_{ex(405, pH\,6.8)} + F_{tot(405, pH\,7.2)} \times x_{in} \tag{5.9}$$

$$F_{tot(460,pH\,6.8)} = F_{ex(460,pH\,6.8)} + F_{tot(460,pH\,7.2)} \times x_{in} \quad (5.10)$$

The ratio of the fluorescence emission at 405 nm and 460 nm (11) at pH 6.8 ($R_{(pH\,6.8)}$) was experimentally determined (see Fig. 1B) and is used to get (12) from (9).

$$R_{(pH\,6.8)} = \frac{F_{ex(405,pH\,6.8)}}{F_{ex(460,pH\,6.8)}} \quad (5.11)$$

$$F_{tot(405,pH\,6.8)} = F_{ex(460,pH\,6.8)} \times R_{(pH\,6.8)} + F_{tot(405,pH\,7.2)} \times x_{in} \quad (5.12)$$

(10) is solved for $F_{a(460,pH\,6.8)}$ and substituted in (12) to yield (13):

$$F_{tot(405,pH\,6.8)} = (F_{tot(460,pH\,6.8)} - F_{tot(460,pH\,7.2)} \times x_{in}) \times R_{(pH6.8)} + F_{tot(405,pH\,7.2)} \times x_{in} \quad (5.13)$$

(13) can be solved after x_{in} to obtain the contribution of internal pyranine.

$$x_{in} = \frac{F_{tot(405,pH\,6.8)} - F_{tot(460,pH\,6.8)} \times R_{(pH\,6.8)}}{F_{tot(405,pH\,7.2)} - F_{tot(460,pH\,7.2)} \times R_{(pH\,6.8)}} \quad (5.14)$$

The contribution of externally bound pyranine can be obtained by (15) for any given pH Y.

$$F_{ex\,(405,pH\,Y)} = F_{tot\,(405,pH\,Y)} - F_{tot\,(405,pH\,7.2)} \times x_{in} \quad (5.15)$$

$$F_{ex\,(460,pH\,Y)} = \frac{F_{ex\,(405,pH\,Y)}}{R_{pH\,Y}} = \frac{F_{tot\,(405,pH\,Y)} - F_{tot\,(405,pH\,7.2)} \times x_{in}}{R_{pH\,Y}} \quad (5.16)$$

Appendix II

Rate calculation for *E. coli* F_0 part in hydrolysis direction, energized with a $\Delta\psi$ of 120 mV

In order to calculate the absolute H^+-transport rate per F_0 molecule, the exact number of transported protons and of F_0 molecules per experiment have to be determined.
First, the total amount of present F_0 parts per measurement was calculated from Fig. 1A. A sample of 10 µl of liposome solution was used per measurement, corresponding to 66 ng of F_1F_0 ATP synthase. If a molecular weight of 550 kDa is assumed for the F_1F_0 ATP synthase, an amount of $1.23 \cdot 10^{-13}$ mol or a total number of $7.224 \cdot 10^{10}$ F_0 parts was present per measurement. Second, the total amount of transported protons is calculated from the pH within the liposomes during the measurement. To obtain the maximal rate at the beginning of the reaction, we fitted the transport process according to Figure 2D and determined the initial slope. For the *E. coli* enzyme, we obtained an initial rate of 0.1432 pH units/ s at pH 7.2 in liposomes containing an F_0 part.
The major contribution of the buffering capacity was made by the phospholipid headgroups inside of the liposomes. From Figure 1C, the buffer capacity of the external layer of liposomes was determined to be 102 µM/ g lipid/ pH unit around the starting pH of the reaction (pH 7.2). The buffering capacity of the internal layer of phospholipids was assumed to be same as the external (Brune et al. 1987). We used the total amount of lipid per measurement (38 µg) and the fraction of liposomes containing an F_0 part (50 %) to calculate the amount of liposomes contributing to the buffer capacity. The molar amount of transported protons is then

$$102 \cdot 10^{-6} \cdot \frac{M}{\text{pH-unit}} \times 14 \cdot 10^{-6} \cdot g \times 0.1432 \cdot \text{pH-unit} = 2.78 \cdot 10^{-10} \cdot M\ H^+.$$

Multiplied with the Avogadro constant N_A, this gives a rate of $1.67 \cdot 10^{14}$ H^+/ s, corresponding to an initial H^+-transport rate of 2350 H^+/ (s × F_0 part).
The buffer capacity of the internal buffer (2 mM MOPS, 1 mM pyranine) can optionally be calculated from the Henderson-Hasselbalch equation for weak acids. Since MOPS and pyranine have similar pKa values ($\sim$ 7.2), their contributions were combined to one internal buffer (3 mM total). The initial pH change (0.1432 pH units/ s) corresponds to a concentration change Δc of 106 µmol $H^+/(l \times s)$ was obtained. Using the internal volume of proteoliposomes per measurement (10 µl liposomes contain 38 µg lipid and possess an internal volume of 59 nl) and considering only the fraction of liposomes containing an F_0 part (50 % in *E. coli*), a number of $1.9 \cdot 10^{12}$ H^+/ s per measurement is obtained, corresponding to 26 H^+/ (s × F_0 part). If the

maximal internal volume (6 µl/ mg lipid) is used for calculation, a number of 102 H^+/ (s × F_0 part) is obtained. The contribution of the internal buffer to the total buffering capacity is not exceeding 5 % and the calculation is therefore nearly independent of the internal volume.

Bibliography

J. P. Abrahams, S. K. Buchanan, M. J. Van Raaij, I. M. Fearnley, A. G. Leslie, and J. E. Walker. The structure of bovine F_1-ATPase complexed with the peptide antibiotic efrapeptin. *Proc. Natl. Acad. Sci. USA*, 93:9420–4, 1996.

J. P. Abrahams, A. G. Leslie, R. Lutter, and J. E. Walker. Structure at 2.8 Å resolution of F_1-ATPase from bovine heart mitochondria. *Nature*, 370:621–8, 1994.

K. Adachi, K. Oiwa, T. Nishizaka, S. Furuike, H. Noji, H. Itoh, M. Yoshida, and Jr. Kinosita, K. Coupling of rotation and catalysis in F_1-ATPase revealed by single-molecule imaging and manipulation. *Cell*, 130:309–21, 2007.

R. Aggeler, M. A. Haughton, and R. A. Capaldi. Disulfide bond formation between the COOH-terminal domain of the b subunits and the γ- and ϵ-subunits of the *Escherichia coli* F_1-ATPase. Structural implications and functional consequences. *J. Biol. Chem.*, 270:9185–91, 1995.

A. Aksimentiev, I. A. Balabin, R. H. Fillingame, and K. Schulten. Insights into the molecular mechanism of rotation in the F_0 sector of ATP synthase. *Biophys. J.*, 86:1332–44, 2004.

C. M. Angevine and R. H. Fillingame. Aqueous access channels in subunit a of rotary ATP synthase. *J. Biol. Chem.*, 278:6066–74, 2003.

C. M. Angevine, K. A. Herold, and R. H. Fillingame. Aqueous access pathways in subunit a of rotary ATP synthase extend to both sides of the membrane. *Proc. Natl. Acad. Sci. USA*, 100:13179–83, 2003.

C. M. Angevine, K. A. Herold, O. D. Vincent, and R. H. Fillingame. Aqueous access pathways in ATP synthase subunit a: Reactivity of cysteine substituted into transmembrane helices 1, 3 and 5. *J. Biol. Chem.*, 282:9001–7, 2007.

R. Birkenhager, M. Hoppert, G. Deckers-Hebestreit, F. Mayer, and K. Altendorf. The F_0 complex of the *Escherichia coli* ATP synthase. Investigation by electron spectroscopic imaging and immunoelectron microscopy. *Eur. J. Biochem.*, 230:58–67, 1995.

Bibliography

C. Bjørbaek, V. Foërsom, and O. Michelsen. The transmembrane topology of the a subunit from the ATPase in *Escherichia coli* analyzed by PhoA protein fusions. *FEBS Lett*, 260: 31–4, 1990.

D. J. Blum, Y. H. Ko, S. Hong, D. A. Rini, and P. L. Pedersen. ATP synthase motor components: proposal and animation of two dynamic models for stator function. *Biochem. Biophys. Res. Commun.*, 287:801–7, 2001.

R. A. Bockmann and H. Grubmüller. Nanoseconds molecular dynamics simulation of primary mechanical energy transfer steps in F_1-ATP synthase. *Nat. Struct. Biol.*, 9:198–202, 2002.

M. W. Bowler, M. G. Montgomery, A. G. Leslie, and J. E. Walker. Ground state structure of F_1-ATPase from bovine heart mitochondria at 1.9 Å resolution. *J. Biol. Chem.*, 282: 14238–42, 2007.

P. D. Boyer. The ATP synthase - a splendid molecular machine. *Annu. Rev. Biochem.*, 66: 717–49, 1997.

K. Braig, R. I. Menz, M. G. Montgomery, A. G. Leslie, and J. E. Walker. Structure of bovine mitochondrial F_1-ATPase inhibited by Mg^{2+} ADP and aluminium fluoride. *Structure*, 8: 567–73, 2000.

M. Branden, T. Sanden, P. Brzezinski, and J. Widengren. Localized proton microcircuits at the biological membrane-water interface. *Proc. Natl. Acad. Sci. USA*, 103:19766–70, 2006.

A. Brune, J. Spillecke, and A. Kröger. Correlation of the turnover number of the ATP synthase in liposomes with the proton flux and the proton potential across the membrane. *Biochim. Biophys. Acta*, 893:499–507, 1987.

E. Cabezon, P. J. Butler, M. J. Runswick, and J. E. Walker. Modulation of the oligomerization state of the bovine F_1-ATPase inhibitor protein, IF_1, by pH. *J. Biol. Chem.*, 275:25460–4, 2000.

E. Cabezon, M. G. Montgomery, A. G. Leslie, and J. E. Walker. The structure of bovine F_1-ATPase in complex with its regulatory protein IF_1. *Nat. Struct. Biol.*, 10:744–50, 2003.

B. D. Cain and R. D. Simoni. Impaired proton conductivity resulting from mutations in the a subunit of F_1F_0 ATPase in *Escherichia coli*. *J. Biol. Chem.*, 261:10043–50, 1986.

B. D. Cain and R. D. Simoni. Interaction between Glu-219 and His-245 within the a-subunit of F_1F_0-ATPase in *Escherichia coli*. *J. Biol. Chem.*, 263:6606–12, 1988.

B. D. Cain and R. D. Simoni. Proton translocation by the F_1F_0 ATPase of *Escherichia coli*. Mutagenic analysis of the a-subunit. *J. Biol. Chem.*, 264:3292–300, 1989.

N. J. Cao, W. S. Brusilow, J. J. Tomashek, and D. J. Woodbury. Characterization of reconstituted F_0-ATPase from wild-type *Escherichia coli* and identification of two other fluxes co-purifying with F_0. *Cell. Biochem. Biophys.*, 34:305–20, 2001.

Y. Cao, G. Váró, A. L. Klinger, D. M. Czajkowsky, M. S. Braiman, R. Needleman, and J. K. Lanyi. Proton transfer from Asp-96 to the bacteriorhodopsin Schiff base is caused by a decrease of the pK_a of Asp-96 which follows a protein backbone conformational change. *Biochemistry*, 32:1981–90, 1993.

R. A. Capaldi and R. Aggeler. Mechanism of the F_1F_0-type ATP synthase, a biological rotary motor. *Trends. Biochem. Sci.*, 27:154–60, 2002.

D. A. Cherepanov, B. A. Feniouk, W. Junge, and A. Y. Mulkidjanian. Low dielectric permittivity of water at the membrane interface: effect on the energy coupling mechanism in biological membranes. *Biophys. J.*, 85:1307–16, 2003.

D. A. Cherepanov and W. Junge. Viscoelastic dynamics of actin filaments coupled to rotary F-ATPase: curvature as an indicator of the torque. *Biophys. J.*, 81:1234–44, 2001.

G. Deckers-Hebestreit and K. Altendorf. The F_1F_0-type ATP synthase of bacteria: Structure and funcion of the F_0 complex. *Annu. Rev. Microbiol.*, 50:791–824, 1996.

Jessica DeLeon-Rangel, Di Zhang, and Steven B Vik. The role of transmembrane span 2 in the structure and function of subunit a of the ATP synthase from *Escherichia coli*. *Arch. Biochem. Biophys.*, 418:55–62, 2003.

N. A. Dencher, P. A. Burghaus, and S. Grzesiek. Determination of the net proton-hydroxide ion permeability across vesicular lipid bilayers and membrane proteins by optical probes. *Methods Enzymol.*, 127:746–60, 1986.

V. K. Dickson, J. A. Silvester, I. M. Fearnley, A. G. Leslie, and J. E. Walker. On the structure of the stator of the mitochondrial ATP synthase. *EMBO J.*, 25:2911–8, 2006.

M. Diez, B. Zimmermann, M. Börsch, M. Konig, E. Schweinberger, S. Steigmiller, R. Reuter, S. Felekyan, V. Kudryavtsev, C. A. Seidel, and P. Graber. Proton-powered subunit rotation in single membrane-bound F_1F_0-ATP synthase. *Nat. Struct. Mol. Biol.*, 11:135–41, 2004.

Bibliography

J. G. Digel, N. D. Moore, and R. E. McCarty. Influence of divalent cations on nucleotide exchange and ATPase activity of chloroplast coupling factor 1. *Biochemistry*, 37:17209–15, 1998.

P. Dimroth. Primary sodium ion translocating enzymes. *Biochim. Biophys. Acta*, 1318:11–51, 1997.

P. Dimroth and G. M. Cook. Bacterial Na^+- or H^+-coupled ATP synthases operating at low electrochemical potential. *Adv. Microb. Physiol.*, 49:175–218, 2004.

P. Dimroth, U. Matthey, and G. Kaim. Critical evaluation of the one- versus the two-channel model for the operation of the ATP synthase's F_0 motor. *Biochim. Biophys. Acta*, 1459:506–13, 2000.

P. Dimroth, C. von Ballmoos, and T. Meier. Catalytic and mechanical cycles in F-ATP synthases. Fourth in the Cycles Review Series. *EMBO Rep.*, 7:276–82, 2006.

O. Dmitriev, G. Deckers-Hebestreit, and K. Altendorf. ATP synthesis energized by ΔpNa and $\Delta\psi$ in proteoliposomes containing the F_1F_0-ATPase from *Propionigenium modestum*. *J. Biol. Chem.*, 226:14776–80, 1993.

O. Dmitriev, P. C. Jones, W. Jiang, and R. H. Fillingame. Structure of the membrane domain of subunit b of the *Escherichia coli* F_1F_0 ATP synthase. *J. Biol. Chem.*, 274:15598–604, 1999.

O.Y Dmitriev, K.H. Freedman, Hermolin J., and R.H. Fillingame. Interaction of transmembrane helices in ATP synthase subunit a in solution as revealed by spin label difference NMR. *Biochim. Biophys. Acta*, 1777:227–37, 2008.

S. Dröse, A. Galkin, and U. Brandt. Proton pumping by complex I (NADH:ubiquinone oxidoreductase) from *Yarrowia lipolytica* reconstituted into proteoliposomes. *Biochim. Biophys. Acta*, 1710:87–95, 2005.

T. M. Duncan, V. V. Bulygin, Y. Zhou, M. L. Hutcheon, and R. L. Cross. Rotation of subunits during catalysis by *Escherichia coli* F_1-ATPase. *Proc. Natl. Acad. Sci. USA*, 92:10964–68, 1995.

S. D. Dunn, D. T. McLachlin, and M. Revington. The second stalk of *Escherichia coli* ATP synthase. *Biochim. Biophys. Acta*, 1458:356–63, 2000.

T. Elston, H. Wang, and G. Oster. Energy transduction in ATP synthase. *Nature*, 391:510–13, 1998.

S. Engelbrecht, G. Deckers Hebestreit, K. Altendorf, and W. Junge. Cross-reconstitution of the F_1F_0-ATP synthases of chloroplasts and *Escherichia coli* with special emphasis on subunit delta. *Eur. J. Biochem.*, 181:485–91, 1989.

C. Etzold, G. Deckers-Hebestreit, and K. Altendorf. Turnover number of *Escherichia coli* F_1F_0 ATP synthase for ATP synthesis in membrane vesicles. *Eur. J. Biochem.*, 243:336–43, 1997.

S. Eya, M. Maeda, and M. Futai. Role of the carboxy terminal region of H^+-ATPase (F_1F_0) a subunit from *Escherichia coli*. *Arch. Biochem. Biophys.*, 284:71–7, 1991.

S. Eya, T. Noumi, M. Maeda, and M. Futai. Intrinsic membrane sector (F_0) of H^+-ATPase (F_1F_0) from *Escherichia coli*. Mutations in the a-subunit give F_0 with impaired proton translocation and F_1 binding. *J. Biol. Chem.*, 263:10056–62, 1988.

B. A. Feniouk, M. A. Kozlova, D. A. Knorre, D. A. Cherepanov, A. Y. Mulkidjanian, and W. Junge. The proton-driven rotor of ATP synthase: ohmic conductance (10 fS), and absence of voltage gating. *Biophys. J.*, 86:4094–109, 2004.

B. A. Feniouk, A. Y. Mulkidjanian, and W. Junge. Proton slip in the ATP synthase of *Rhodobacter capsulatus*: induction, proton conduction, and nucleotide dependence. *Biochim. Biophys. Acta*, 1706:184–94, 2005.

R. H. Fillingame. Molecular mechanism of ATP synthesis by F_1F_0- Type H^+-transporting ATP synthases. In *The Bacteria*, volume XII, pages 345–391. 1990.

R. H. Fillingame, W. Jiang, and O. Y. Dmitriev. Coupling H^+ transport to rotary catalysis in F-type ATP synthases: structure and organization of the transmembrane rotary motor. *J. Exp. Biol.*, 203:9–17, 2000.

S. Fischer and P. Gräber. Comparison of ΔpH- and $\Delta\psi$-driven ATP synthesis catalyzed by the H^+-ATPases from *Escherichia coli* or chloroplasts reconstituted into liposomes. *FEBS Lett*, 457:327–32, 1999.

S. Fischer, P. Gräber, and P. Turina. The activity of the ATP synthase from *Escherichia coli* is regulated by the transmembrane proton motive force. *J. Biol. Chem.*, 275:30157–62, 2000.

M. J. Franklin, W. S. Brusilow, and D. J. Woodbury. Determination of proton flux and conductance at pH 6.8 through single F_0 sectors from *Escherichia coli*. *Biophys. J.*, 87:3594–9, 2004.

Bibliography

J. L. Gardner and B. D. Cain. Amino acid substitutions in the a subunit affect the epsilon subunit of F_1F_0 ATP synthase from *Escherichia coli*. *Arch. Biochem. Biophys.*, 361:302–8, 1999.

U. Gerike, G. Kaim, and P. Dimroth. *In vivo* synthesis of ATPase complexes of *Propionigenium modestum* and *Escherichia coli* and analysis of their function. *Eur. J. Biochem.*, 232:596–602, 1995.

E. P. Gogol, E. Johnston, R. Aggeler, and R. A. Capaldi. Ligand-dependent structural variations in *Escherichia coli* F_1 ATPase revealed by cryoelectron microscopy. *Proc. Natl. Acad. Sci. USA*, 87:9585–9, 1990.

E. A. Gorbikova, N. P. Belevich, M. Wikström, and M. I. Verkhovsky. Protolytic reactions on reduction of cytochrome c oxidase studied by ATR-FTIR spectroscopy. *Biochemistry*, 46: 4177–83, 2007.

J. C. Greie, G. Deckers-Hebestreit, and K. Altendorf. Secondary structure composition of reconstituted subunit b of the *Escherichia coli* ATP synthase. *Eur. J. Biochem.*, 267:3040–8, 2000.

J. C. Greie, T. Heitkamp, and K. Altendorf. The transmembrane domain of subunit b of the *Escherichia coli* F_1F_0 ATP synthase is sufficient for H^+-translocating activity together with subunits a and c. *Eur. J. Biochem.*, 271:3036–42, 2004.

A. A. Guffanti, R. T. Fuchs, M. Schneier, E. Chiu, and T. A. Krulwich. A transmembrane electrical potential generated by respiration is not equivalent to a diffusion potential of the same magnitude for ATP synthesis by *Bacillus firmus* RAB. *J. Biol. Chem.*, 259:2971–5, 1984.

P. E. Hartzog and B.D. Cain. Second-site supressor mutations at Glycine 218 and Histidine 245 in the a-subunit of F_1F_0 ATP synthase in *Escherichia coli*. *J. Biol. Chem.*, 269:32313–7, 1994.

P. E. Hartzog, J. L. Gardner, and B. D. Cain. Modeling the Leigh syndrome nt8993 T->C mutation in *Escherichia coli* F_1F_0 ATP synthase. *Int. J. Biochem. Cell Biol.*, 31:769–76, 1999.

L. P. Hatch, G. B. Cox, and S. M. Howitt. The essential arginine residue at position 210 in the a-subunit of the *Escherichia coli* ATP synthase can be transferred to position 252 with partial retention of activity. *J. Biol. Chem.*, 270:29407–12, 1995.

J. Heberle, J. Riesle, G. Thiedemann, D. Oesterhelt, and N. A. Dencher. Proton migration along the membrane surface and retarded surface to bulk transfer. *Nature*, 370:379–82, 1994.

D. B. Hicks, Z. Wang, Y. Wei, R. Kent, A. A. Guffanti, H. Banciu, D. H. Bechhofer, and T. A. Krulwich. A tenth atp gene and the conserved atpI gene of a *Bacillus* atp operon have a role in Mg^{2+} uptake. *Proc. Natl. Acad. Sci. USA*, 100:10213–8, 2003.

W. Hilpert, B. Schink, and P. Dimroth. Life by a new decarboxylation-dependent energy conservation mechanism with Na^+ as coupling ion. *EMBO J.*, 3:1665–70, 1984.

J. Hoppe, H. U. Schairer, P. Friedl, and W. Sebald. An Asp-Asn substitution in the proteolipid subunit of the ATP-synthase from *Escherichia coli* leads to a non-functional proton channel. *FEBS Lett.*, 145:21–9, 1982.

S. M. Howitt, F. Gibson, and G. B. Cox. The proton pore of the F_1F_0-ATPase of *Escherichia coli*: Ser-206 is not required for proton translocation. *Biochim. Biophys. Acta*, 936:74–80, 1988.

S. M. Howitt, R. N. Lightowlers, F. Gibson, and G. B. Cox. Mutational analysis of the function of the a-subunit of the F_1F_0-ATPase of *Escherichia coli*. *Biochim. Biophys. Acta*, 1015:264–268, 1990.

S. M. Howitt, A. J. Rodgers, L. P. Hatch, F. Gibson, and G. B. Cox. The coupling of the relative movement of the a and c subunits of the F_0 to the conformational changes in the F_1-ATPase. *J. Bioenerg. Biomembr.*, 28(5):415–420., 1996.

C. Hunte, E. Screpanti, M. Venturi, A. Rimon, E. Padan, and H. Michel. Structure of a Na^+/H^+ antiporter and insights into mechanism of action and regulation by pH. *Nature*, 435:1197–202, 2005.

M. L. Hutcheon, T. M. Duncan, H. Ngai, and R. L. Cross. Energy-driven subunit rotation at the interface between subunit a and the c oligomer in the F_0 sector of *Escherichia coli* ATP synthase. *Proc. Natl. Acad. Sci. USA*, 98:8519–24., 2001.

R. R. Ishmukhametov, M. A. Galkin, and S. B. Vik. Ultrafast purification and reconstitution of His-tagged cysteine-less *Escherichia coli* F_1F_0 ATP synthase. *Biochim. Biophys. Acta*, 1706:110–6, 2005.

D. M. Ivey and T. A. Krulwich. Two unrelated alkaliphilic *Bacillus* species possess identical deviations in sequence from those of other prokaryotes in regions of F_0-ATPase proposed to

Bibliography

be involved in proton translocation through the ATP synthase. *Res. Microbiol.*, 143:467–70, 1992.

A. T. Jagendorf and E. Uribe. ATP formation caused by acid-base transition of spinach chloroplasts. *Proc. Natl. Acad. Sci. USA*, 55:170–7, 1966.

H. Jäger, R. Birkenhäger, W-D. Stalz, K. Altendorf, and G. Deckers-Hebestreit. Topology of subunit a of the *Escherichia coli* ATP synthase. *Eur. J. Biochem.*, 251:122–32, 1998.

W. Jiang and R. H. Fillingame. Interacting helical faces of subunits a and c in the F_1F_0 ATP synthase of *Escherichia coli* defined by disulfide cross-linking. *Proc. Natl. Acad. Sci. USA*, 95:6607–12, 1998.

E. A. Johnson and R. E. McCarty. The carboxyl terminus of the epsilon subunit of the chloroplast ATP synthase is exposed during illumination. *Biochemistry*, 41:2446–51, 2002.

W. Junge, H. Lill, and S. Engelbrecht. ATP synthase: an electrochemical transducer with rotatory mechanics. *Trends Biochem. Sci.*, 22:420–3, 1997.

R. Kagawa, M. G. Montgomery, K. Braig, A. G. Leslie, and J. E. Walker. The structure of bovine F_1-ATPase inhibited by ADP and beryllium fluoride. *EMBO J.*, 23:2734–44, 2004.

G. Kaim and P. Dimroth. Formation of a functionally active sodium-translocating hybrid F_1F_0 ATPase in *Escherichia coli* by homologous recombination. *Eur. J. Biochem.*, 218:937–44, 1993.

G. Kaim and P. Dimroth. Construction, expression and characterization of a plasmid-encoded Na^+-specific ATPase hybrid consisting of *Propionigenium modestum* F_0-ATPase and *Escherichia coli* F_1-ATPase. *Eur. J. Biochem.*, 222:615–23, 1994.

G. Kaim and P. Dimroth. ATP synthesis by F-type ATP synthase is obligatorily dependent on the transmembrane voltage. *EMBO J.*, 18:4118–27., 1999.

G. Kaim, U. Matthey, and P. Dimroth. Mode of interaction of the single a-subunit with the multimeric c-subunits during the translocation of the coupling ions by F_1F_0 ATPases. *EMBO J.*, 17:688–95, 1998.

G. Kaim, F. Wehrle, U. Gerike, and P. Dimroth. Molecular basis for the coupling ion selectivity of F_1F_0 ATP synthases: probing the liganding groups for Na^+ and Li^+ in the c-subunit of the ATP synthase from *Propionigenium modestum*. *Biochemistry*, 36:9185–94, 1997.

Y. Kato, T. Matsui, N. Tanaka, E. Muneyuki, T. Hisabori, and M. Yoshida. Thermophilic F_1-ATPase is activated without dissociation of an endogenous inhibitor, epsilon subunit. *J. Biol. Chem.*, 272:24906–12, 1997.

Y. Kato-Yamada, D. Bald, M. Koike, K. Motohashi, T. Hisabori, and M. Yoshida. Epsilon subunit, an endogenous inhibitor of bacterial F_1-ATPase, also inhibits F_1F_0-ATPase. *J. Biol. Chem.*, 274:33991–4, 1999.

Y. Kato-Yamada and M. Yoshida. Isolated epsilon subunit of thermophilic F_1-ATPase binds ATP. *J. Biol. Chem.*, 278:36013–6, 2003.

C. Kayalar, J. Rosing, and P. D. Boyer. An alternating site sequence for oxidative phosphorylation suggested by measurement of substrate binding patterns and exchange reaction inhibitions. *J. Biol. Chem.*, 252:2486–91, 1977.

D. J. Klionsky, W. S. Brusilow, and R. D. Simoni. *In vivo* evidence for the role of the epsilon subunit as an inhibitor of the proton-translocating ATPase of *Escherichia coli*. *J. Bacteriol.*, 160:1055–60, 1984.

C. Kluge and P. Dimroth. Studies on Na^+ and H^+ translocation through the F_0 part of the Na^+- translocating F_1F_0 ATPase from *Propionigenium modestum*: discovery of a membrane potential dependent step. *Biochemistry*, 31:12665–72, 1992.

C. Kluge and P. Dimroth. Kinetics of inactivation of the F_1F_0 ATPase of *Propionigenium modestum* by dicyclohexylcarbodiimide in relationship to H^+ and Na^+ concentration: probing the binding site for the coupling ions. *Biochemistry*, 32:10378–86, 1993a.

C. Kluge and P. Dimroth. Specific protection by Na^+ or Li^+ of the F_1F_0-ATPase of *Propionigenium modestum* from the reaction with dicyclohexylcarbodiimide. *J. Biol. Chem.*, 268:14557–60, 1993b.

T. A. Krulwich. Alkaliphiles: 'basic' molecular problems of pH tolerance and bioenergetics. *Mol. Microbiol.*, 15:403–10, 1995.

T. A. Krulwich and A. A. Guffanti. Proton-coupled bioenergetic processes in extremely alkaliphilic bacteria. *J. Bioenerg. Biomembr.*, 24:587–99, 1992.

J. Kríz, E: Makrlík, and P. Vanura. NMR evidence of a valinomycin-proton complex. *Biopolymers*, 81:104–9, 2006.

Bibliography

C. A. Kumamoto and R. D. Simoni. Genetic evidence for interaction between the a-subunits and b-subunits of the F_0 portion of the *Escherichia coli* proton translocating ATPase. *J. Biol. Chem.*, 261:10037–42, 1986.

P. H. Kuo and R. K. Nakamoto. Intragenic and intergenic suppression of the *Escherichia coli* ATP synthase subunit a mutation of Gly-213 to Asn: functional interactions between residues in the proton transport site. *Biochem. J.*, 347:797–805, 2000.

L. Langemeyer and S. Engelbrecht. Essential arginine in subunit a and aspartate in subunit c of F_1F_0 ATP synthase. Effect of repositioning within Helix 4 of subunit a and Helix 2 of subunit c. *Biochim. Biophys. Acta*, 1767:998–1005, 2007.

W. Laubinger and P. Dimroth. Characterization of the Na^+-stimulated ATPase of *Propionigenium modestum* as an enzyme of the F_1F_0 type. *Eur. J. Biochem.*, 168:475–80, 1987.

W. Laubinger and P. Dimroth. Characterization of the ATP synthase of *Propionigenium modestum* as a primary sodium pump. *Biochemistry*, 27:7531–7, 1988.

W. Laubinger and P. Dimroth. The sodium ion translocating adenosinetriphosphatase of *Propionigenium modestum* pumps protons at low sodium ion concentrations. *Biochemistry*, 28:7194–8, 1989.

M. J. Lewis, J. A. Chang, and R. D. Simoni. A topological analysis of subunit a from *Escherichia coli* F_1F_0-ATP synthase predicts eight transmembrane segments. *J. Biol. Chem.*, 265:10541–50, 1990.

R. N. Lightowlers, S. M. Howitt, L. Hatch, F. Gibson, and G. Cox. The proton pore in the *Escherichia coli* F_1F_0-ATPase: Substitution of glutamate by glutamine at position 219 of the a-subunit prevents F_0-mediated proton permeability. *Biochim. Biophys. Acta*, 933:241–8, 1988.

R. N. Lightowlers, S. M. Howitt, L. Hatch, F. Gibson, and G. B. Cox. The proton pore in *Escherichia coli* F_1F_0- ATPase: A requirement of arginine at position 210 of the a-subunit. *Biochim. Biophys. Acta*, 894:399–406, 1987.

H. Lill, S. Engelbrecht, G. Schonknecht, and W. Junge. The proton channel, CF_0, in thylakoid membranes. Only a low proportion of CF_1-lacking CF_0 is active with a high unit conductance (169 fS). *Eur. J. Biochem.*, 160:627–34, 1986.

J. C. Long, J. DeLeon-Rangel, and S. B. Vik. Characterization of the first cytoplasmic loop of subunit a of the *Escherichia coli* ATP synthase by surface labeling, cross-linking, and mutagenesis. *J. Biol. Chem.*, 277:27288–93, 2002.

J. C. Long, S. Wang, and S. B. Vik. Membrane topology of subunit a of the F_1F_0 ATP synthase as determined by labelling of unique cysteine residues. *J. Biol. Chem.*, 273:16235–40, 1998.

A. Matin, E. Zychlinsky, M. Keyhan, and G. Sachs. Capacity of *Helicobacter pylori* to generate ionic gradients at low pH is similar to that of bacteria which grow under strongly acidic conditions. *Infect. Immun.*, 64:1434–36, 1996.

D. G. McMillan, S. Keis, P. Dimroth, and G. M. Cook. A specific adaptation in the a subunit of thermoalkaliphilic F_1F_0-ATP synthase enables ATP synthesis at high pH but not at neutral pH values. *J. Biol. Chem.*, 282:17395–404, 2007.

T. Meier and P. Dimroth. Intersubunit bridging by Na^+ ions as a rationale for the unusual stability of the c-rings of Na^+-translocating F_1F_0 ATP synthases. *EMBO Rep.*, 3:1094–8, 2002.

T. Meier, U. Matthey, C. von Ballmoos, J. Vonck, T. Krug von Nidda, W. Kühlbrandt, and P. Dimroth. Evidence for structural integrity in the undecameric c-rings isolated from sodium ATP synthases. *J. Mol. Biol.*, 325:389–97, 2003.

T. Meier, N. Morgner, D. Matthies, D. Pogoryelov, S. Keis, G. M. Cook, P. Dimroth, and B. Brutschy. A tridecameric c ring of the adenosine triphosphate (ATP) synthase from the thermoalkaliphilic *Bacillus* sp. strain TA2.A1 facilitates ATP synthesis at low electrochemical proton potential. *Mol. Microbiol.*, 65:1181–92, 2007.

T. Meier, P. Polzer, K. Diederichs, W. Welte, and P. Dimroth. Structure of the rotor ring of F-type Na^+-ATPase from *Ilyobacter tartaricus*. *Science*, 308:659–62, 2005.

R. I. Menz, J. E. Walker, and A. G. Leslie. Structure of bovine mitochondrial F_1-ATPase with nucleotide bound to all three catalytic sites: implications for the mechanism of rotary catalysis. *Cell*, 106:331–41, 2001.

H. Michel and D. Oesterhelt. Electrochemical proton gradient across the cell membrane of *Halobacterium halobium*: comparison of the light-induced increase with the increase of intracellular adenosine triphosphate under steady-state illumination. *Biochemistry*, 19:4615–9, 1980a.

Bibliography

H. Michel and D. Oesterhelt. Electrochemical proton gradient across the cell membrane of *Halobacterium halobium*: effect of N,N'-dicyclohexylcarbodiimide, relation to intracellular adenosine triphosphate, adenosine diphosphate, and phosphate concentration, and influence of the potassium gradient. *Biochemistry*, 19:4607–14, 1980b.

P. Mitchell. Keilin's respiratory chain concept and its chemiosmotic consequences. *Science*, 206:1148–59, 1979.

K. J. Moore, C. M. Angevine, Vincent. O. D., B. E. Schwem, and R. H. Fillingame. The cytoplasmic loops of subunit a of Escherichia coli ATP synthase may participate in the proton translocating mechanism. *J. Biol. Chem.*, 283:13044–52, 2008.

Y. Moriyama, A. Iwamoto, H. Hanada, M. Maeda, and M. Futai. One-step purification of *Escherichia coli* H^+-ATPase (F_1F_0) and its reconstitution into liposomes with neurotransmitter transporters. *J. Biol. Chem.*, 266:22141–6, 1991.

T. Murata, K. Igarashi, Y. Kakinuma, and I. Yamato. Na^+ binding of V-type Na^+-ATPase in *Enterococcus hirae*. *J. Biol. Chem.*, 275:13415–9, 2000.

T. Murata, I. Yamato, Y. Kakinuma, A. G. Leslie, and J. E. Walker. Structure of the rotor of the V-Type Na^+-ATPase from Enterococcus hirae. *Science*, 308(5722):654–9, 2005.

T. Nakano, T. Ikegami, T. Suzuki, M. Yoshida, and H. Akutsu. A new solution structure of ATP synthase subunit c from thermophilic *Bacillus* PS3, suggesting a local conformational change for H^+-translocation. *J. Mol. Biol.*, 358:132–44, 2006.

C. M. Nalin and R. E. McCarty. Role of a disulfide bond in the gamma subunit in activation of the ATPase of chloroplast coupling factor 1. *J. Biol. Chem.*, 259:7275–80, 1984.

S. Neumann, U. Matthey, G. Kaim, and P. Dimroth. Purification and properties of the F_1F_0 ATPase of *Ilyobacter tartaricus*, a sodium ion pump. *J. Bacteriol.*, 180(13):3312–16, 1998.

H. Noji, R. Yasuda, M. Yoshida, and Jr. Kinosita, K. Direct observation of the rotation of F_1-ATPase. *Nature*, 386:299–302, 1997.

K. F. Nowak and R. E. McCarty. Regulatory role of the C-terminus of the epsilon subunit from the chloroplast ATP synthase. *Biochemistry*, 43:3273–9, 2004.

A. R. Patterson, T. Wada, and S. B. Vik. His_{15} of subunit a of the *Escherichia coli* ATP synthase is important for the structure or assembly of the membrane sector F_0. *Arch. Biochem. Biophys.*, 368:193–7, 1999.

C. R. Paule and R. H. Fillingame. Mutations in three of the putative transmembrane helices of subunit a of the *Escherichia coli* F_1F_0-ATPase disrupt ATP-driven proton translocation. *Arch. Biochem. Biophys.*, 274:270–84, 1989.

H. S. Penefsky, M. E. Pullman, A. Datta, and E. Racker. Partial resolution of the enzymes catalyzing oxidative phosphorylation. II. Participation of a soluble adenosine triphosphatase in oxidative phosphorylation. *J. Biol. Chem.*, 235:3330–6, 1960.

O. Pänke, D. A. Cherepanov, K. Gumbiowski, S. Engelbrecht, and W. Junge. Viscoelastic dynamics of actin filaments coupled to rotary F-ATPase: angular torque profile of the enzyme. *Biophys. J.*, 81:1220–33, 2001.

D. Pogoryelov, J. Yu, T. Meier, J. Vonck, P. Dimroth, and D. J. Müller. The c_{15} ring of the *Spirulina platensis* F-ATP synthase: F_1/F_0 symmetry mismatch is not obligatory. *EMBO Rep.*, 6:1040–4, 2005.

M. E. Pullman, H. S. Penefsky, A. Datta, and E. Racker. Partial resolution of the enzymes catalyzing oxidative phosphorylation. I. Purification and properties of soluble dinitrophenol-stimulated adenosine triphosphatase. *J. Biol. Chem.*, 235:3322–9, 1960.

V. K. Rastogi and M. E. Girvin. ^{1}H, ^{13}C, and ^{15}N assignments and secondary structure of the high pH form of subunit c of the F_1F_0 ATP synthase. *J. Biomol. NMR*, 13:91–2, 1999a.

V. K. Rastogi and M. E. Girvin. Structural changes linked to proton translocation by subunit c of the ATP synthase. *Nature*, 402:263–8, 1999b.

M. Renganathan, E. Pfündel, and Richard A. Dilley. Thylakoid lumenal pH determination using a fluorescent dye: Correlation of lumen pH and gating between localized and delocalized energy coupling. *Biochim. Biophys. Acta*, 1142:277–92, 1993.

M. L. Richter, W. J. Patrie, and R. E. McCarty. Preparation of the epsilon subunit and epsilon subunit-deficient chloroplast coupling factor 1 in reconstitutively active forms. *J. Biol. Chem.*, 259:7371–3, 1984.

E. Schneider and K. Altendorf. All three subunits are required for the reconstitution of an active proton channel (F_0) of *Escherichia coli* ATP synthase (F_1F_0). *EMBO J.*, 4:515–8, 1985.

B. Schulenberg, R. Aggeler, J. Murray, and R. A. Capaldi. The γ-c subunit interface in the ATP synthase of *Escherichia coli*. *J. Biol. Chem.*, 274:34233–7, 1999.

Bibliography

B. E. Schwem and R. H. Fillingame. Cross-linking between helices within subunit a of *Escherichia coli* ATP synthase defines the transmembrane packing of a four-helix bundle. *J. Biol. Chem.*, 281:37861–7, 2006.

A. E. Senior, J. A. Downie, G. B. Cox, F. Gibson, L. Langman, and D. R. Fayle. The uncA gene codes for the alpha-subunit of the adenosine triphosphatase of *Escherichia coli*. Electrophoretic analysis of uncA mutant strains. *Biochem. J.*, 180:103–9, 1979.

N. Sone, T. Hamamoto, and Y. Kagawa. pH dependence of H^+ conduction through the membrane moiety of the H^+-ATPase (F_1F_0) and effects of tyrosyl residue modification. *J. Biol. Chem.*, 256:2873–7, 1981.

N. Sone, M. Yoshida, H. Hirata, and Y. Kagawa. Adenosine triphosphate synthesis by electrochemical proton gradient in vesicles reconstituted from purified adenosine triphosphatase and phospholipids of thermophilic bacterium. *J. Biol. Chem.*, 252:2956–60, 1977.

P. L. Sorgen, M. R. Bubb, K. A. McCromick, A. S. Edison, and B. D. Cain. Formation of the b-subunit dimer is necessary for interaction with F_1-ATPase. *Biochemistry*, 37:923–32, 1998a.

P. L. Sorgen, T. L. Caviston, R. C. Perry, and B. D. Cain. Deletions in the second stalk of the F_1F_0-ATPase in *Escherichia coli*. *J. Biol. Chem.*, 273:27873–8, 1998b.

P. R. Steed and R. H. Fillingame. Subunit a facilitates aqueous access to a membrane-embedded region of subunit c in *Escherichia coli* F_1F_0 ATP synthase. *J. Biol. Chem.*, 283:12365–72, 2008.

D. Stock, A. G. Leslie, and J. E. Walker. Molecular architecture of the rotary motor in ATP synthase. *Science*, 286:1700–5, 1999.

Lubert Stryer. *Biochemistry*. W.H. Freeman and Company, New York, 1995.

T. Suzuki, T. Murakami, R. Iino, J. Suzuki, S. Ono, Y. Shirakihara, and M. Yoshida. F_1F_0-ATPase/synthase is geared to the synthesis mode by conformational rearrangement of epsilon subunit in response to proton motive force and ADP/ATP balance. *J. Biol. Chem.*, 278: 46840–6, 2003.

T. Suzuki, Y. Ozaki, N. Sone, B. A. Feniouk, and M. Yoshida. The product of uncI gene in F_1F_0-ATP synthase operon plays a chaperone-like role to assist c-ring assembly. *Proc. Natl. Acad. Sci. USA*, 104:20776–81, 2007.

Bibliography

T. Tsukihara, H. Aoyama, E. Yamashita, T. Tomizaki, H. Yamaguchi, K. Shinzawa-Itoh, R. Nakashima, R. Yaono, and S. Yoshikawa. The whole structure of the 13-subunit oxidized cytochrome c oxidase at 2.8 Å. *Science*, 272:1136–44, 1996.

S. P. Tsunoda, R. Aggeler, M. Yoshida, and R. A. Capaldi. Rotation of the c subunit oligomer in fully functional F_1F_0 ATP synthase. *Proc. Natl. Acad. Sci. USA*, 98:898–902, 2001a.

S. P. Tsunoda, A. J. Rodgers, R. Aggeler, M. C. Wilce, M. Yoshida, and R. A. Capaldi. Large conformational changes of the ϵ subunit in the bacterial F_1F_0 ATP synthase provide a ratchet action to regulate this rotary motor enzyme. *Proc. Natl. Acad. Sci. USA*, 98:6560–4, 2001b.

P. Turina, D. Samoray, and P. Gräber. H^+/ATP ratio of proton transport-coupled ATP synthesis and hydrolysis catalysed by CF_1F_0-liposomes. *EMBO J.*, 22:418–26, 2003.

U. Uhlin, G. B. Cox, and J. M. Guss. Crystal structure of the epsilon subunit of the proton-translocating ATP synthase from *Escherichia coli*. *Structure*, 5:1219–30, 1997.

F. I. Valiyaveetil and R. H. Fillingame. On the role of Arg-210 and Glu-219 of subunit a in proton translocation by the *Escherichia coli* F_1F_0-ATP synthase. *J. Biol. Chem.*, 272(51): 32635–32641, 1997.

F. I. Valiyaveetil and R. H. Fillingame. Transmembrane topography of subunit a in the *Escherichia coli* F_1F_0 ATP synthase. *J. Biol. Chem.*, 273:16241–7, 1998.

S. B. Vik and B. J. Antonio. A mechanism of proton translocation by F_1F_0 ATP synthases suggested by double mutants of the a-subunit. *J. Biol. Chem.*, 269:30364–9, 1994.

S. B. Vik, B. D. Cain, K. T. Chun, and R. D. Simoni. Mutagenesis of the a-subunit of the F_1F_0-ATPase from Escherichia coli. Mutations at Glu-196, Pro-190, and Ser-199. *J. Biol. Chem.*, 263:6599–605, 1988.

S. B. Vik, D. Lee, C. E. Curtis, and L. T. Nguyen. Mutagenesis of the a-subunit of the F_1F_0-ATP synthase from *Escherichia coli* in the region of Asn-192. *Arch. Biochem. Biophys.*, 282: 125–31, 1990.

S. B. Vik, A. R. Patterson, and B. J. Antonio. Insertion scanning mutagenesis of subunit a of the F_1F_0 ATP synthase near His245 and implications on gating of the proton channel. *J. Biol. Chem.*, 273:16229–34, 1998.

S.B. Vik, D. Lee, and P. A. Marshall. Temperature-sensitive mutations at the carboxy terminus of the a-subunit of the *Escherichia coli* F_1F_0 ATP synthase. *J. Bacteriol.*, 173:4544–8, 1991.

Bibliography

C. von Ballmoos, Y. Appoldt, J. Brunner, T. Granier, A. Vasella, and P. Dimroth. Membrane topography of the coupling ion binding site in Na$^+$-translocating F_1F_0 ATP synthase. *J. Biol. Chem.*, 277:3504–10, 2002a.

C. von Ballmoos, J. Brunner, and P. Dimroth. The ion channel of F-ATP synthase is the target of toxic organotin compounds. *Proc. Natl. Acad. Sci. USA*, 101:11239–44, 2004.

C. von Ballmoos, G. M. Cook, and P. Dimroth. Unique Rotary ATP Synthase and Its Biological Diversity. *Annual Review of Biophysics*, 37:43–64, 2008.

C. von Ballmoos and P. Dimroth. Two distinct proton binding sites in the ATP synthase family. *Biochemistry*, 46:11800–9, 2007.

C. von Ballmoos, T. Meier, and P. Dimroth. Membrane embedded location of Na$^+$ or H$^+$ binding sites on the rotor ring of F_1F_0 ATP synthases. *Eur. J. Biochem.*, 269:5581–9, 2002b.

J. Vonck, T. Krug von Nidda, T. Meier, U. Matthey, D. J. Mills, W. Kühlbrandt, and P. Dimroth. Molecular architecture of the undecameric rotor of a bacterial Na$^+$-ATP synthase. *J. Mol. Biol.*, 321:307–16, 2002.

T. Vorburger, J Zingg Ebneter, A. Wiedenmann, D. Morger, G. Weber, K. Diederichs, P. Dimroth, and C. von Ballmoos. Arginine-induced conformational change in the c-ring/a-subunit interface of ATP synthase. *FEBS J.*, 275:2137–50, 2008.

W. Wada, J. C. Long, D. Zhang, and S. B. Vik. A novel labeling approach supports the five-transmembrane model of subunit a of the *Escherichia coli* ATP synthase. *J. Biol. Chem.*, 274:17353–7, 1999.

John E. Walker. ATP Synthesis by Rotary Catalysis (Nobel lecture). *Angew. Chem. Int. Ed.*, 37:2308–19, 1998.

Z. Wang, D. B. Hicks, A. A. Guffanti, K. Baldwin, and T. A. Krulwich. Replacement of amino acid sequence features of a- and c-subunits of ATP synthases of alkaliphilic *Bacillus* with the *Bacillus* consensus sequence results in defective oxidative phosphorylation and non-fermentative growth at pH 10.5. *J. Biol. Chem.*, 279:26546–54, 2004.

F. Wehrle, G. Kaim, and P. Dimroth. Molecular mechanism of the ATP synthase's F_0 motor probed by mutational analyses of subunit a. *J. Mol. Biol.*, 322:369–81, 2002.

A. Wiedenmann, P. Dimroth, , and C. Von Ballmoos. $\Delta\psi$ and ΔpH are equivalent driving forces for proton transport through isolated F_0 complexes of ATP synthases. *Biochim. Biophys. Acta*, 1777:1301–10, 2008.

Bibliography

M. Wikström and M.I. Verkhovsky. Mechanism and energetics of proton translocation by the respiratory heme-copper oxidases. *Biochim. Biophys. Acta*, 1767:1200–14, 2007.

S. Wilkens. Solution structure of the ϵ subunit of the F_1-ATPase from *Escherichia coli* and interactions of this subunit with b-subunits in the complex. *J. Biol. Chem.*, 273:26645–51, 1998.

R. J. Williams. The multifarious couplings of energy transduction. *Biochim. Biophys. Acta*, 505:1–44, 1978.

J. Xing, H. Wang, C. Von Ballmoos, P. Dimroth, and G. Oster. Torque Generation by the F_0 motor of the Sodium ATPase. *Biophys. J.*, 87:2148–63, 2004.

H. Yagi, N. Kajiwara, H. Tanaka, T. Tsukihara, Y. Kato-Yamada, M. Yoshida, and H. Akutsu. Structures of the thermophilic F_1-ATPase epsilon subunit suggesting ATP-regulated arm motion of its C-terminal domain in F_1. *Proc. Natl. Acad. Sci. USA*, 104:11233–8, 2007.

H. Yamada, Y. Moriyama, M. Maeda, and M. Futai. Transmembrane topology of *Escherichia coli* H^+-ATPase (ATP synthase) subunit a. *FEBS Lett.*, 390:34–8, 1996.

R. Yasuda, H. Noji, K. Kinosita, and M. Yoshida. F_1-ATPase is a highly efficient molecular motor that rotates with discrete 120 degree steps. *Cell*, 93:1117–24, 1998.

R. Yasuda, H. Noji, M. Yoshida, Jr. Kinosita, K., and H. Itoh. Resolution of distinct rotational substeps by submillisecond kinetic analysis of F_1-ATPase. *Nature*, 410:898–904, 2001.

T. V. Zharova and A. D. Vinogradov. Proton-translocating ATP-synthase of *Paracoccus denitrificans*: ATP-hydrolytic activity. *Biochemistry (Mosc)*, 68:1101–8, 2003.

VDM Verlagsservicegesellschaft mbH

Die VDM Verlagsservicegesellschaft sucht für wissenschaftliche Verlage abgeschlossene und herausragende

Dissertationen, Habilitationen, Diplomarbeiten, Master Theses, Magisterarbeiten usw.

für die kostenlose Publikation als Fachbuch.

Sie verfügen über eine Arbeit, die hohen inhaltlichen und formalen Ansprüchen genügt, und haben Interesse an einer honorarvergüteten Publikation?

Dann senden Sie bitte erste Informationen über sich und Ihre Arbeit per Email an *info@vdm-vsg.de*.

Sie erhalten kurzfristig unser Feedback!

VDM Verlagsservicegesellschaft mbH
Dudweiler Landstr. 99
D - 66123 Saarbrücken
Telefon +49 681 3720 174
Fax +49 681 3720 1749
www.vdm-vsg.de

Die VDM Verlagsservicegesellschaft mbH vertritt

Printed by Books on Demand GmbH, Norderstedt / Germany